ELECTRONIC DESIGN AUTOMATION OF MULTI-SCROLL CHAOS GENERATORS

By

Jesus Manuel Muñoz Pacheco and Esteban Tlelo Cuautle

eBooks End User License Agreement

CONTENTS

DEDICATION

I want to dedicate this book to my future wife, Lízbeth, without her encouragement, love, and valuable advice, write this book would not have been possible.

It is difficult to write a technical book without contributions from others. For their generous help, I want to thank Dr. Carlos Sánchez López, Rodolfo Trejo Guerra, Victor Hugo Carbajal Gómez and Erika Perez Aguilar, who played a big part in helping me with this first edition.

Finally, I thank CONACyT/MEXICO and INAOE for their financial support granted through my Ph.D. scholarship #204409 during 2006-2009 and the facilities during the realization of this Book.

Jesus Manuel

To all my family, to my father Felipe and my mother Gabriela, with special dedication to Maria Alberta, Itzel, Esteban, Aydeé and Yaretzi for the time I used to write this book instead of sharing it with them.

To CONACyT and INAOE for the funding through the project 48396-Y and my sabbatical leave at University of California Riverside during 2009-2010.

Esteban

AFFILIATIONS

Jesus Manuel Muñoz Pacheco

INAOE (National Institute for Astrophysics, Optics and Electronics), Department of Electronics. www.inaoe.gob.mx

Esteban Tlelo Cuautle

INAOE (National Institute for Astrophysics, Optics and Electronics), Department of Electronics. www.inaoe.gob.mx

ABOUT THE AUTHORS

Jesus Manuel Muñoz-Pacheco got the B.Sc. degree from Instituto Tecnológico de Orizaba (ITO), México, in 2003 and his M.Sc. and Ph.D. degrees from Instituto Nacional de Astrofísica, Óptica y Elctrónica (INAOE), México, in 2005 and 2009, respectively. Between 2008 and 2009 he was Associate Professor at the Dept. of Electronics and Telecommunications of Universidad Politécnica de Puebla, México, and since 2009 he is Full Professor-Researcher at the same University. His research interests lie in the behavioural modelling and simulation of linear and nonlinear circuits and systems, chaotic oscillators and analog design automation tools. Dr. Muñoz-Pacheco has authored and co-authored one book, two book chapters and about 20 papers in international journals and conferences. He has also participated as researcher or main researcher in Research Projects for National Council of Science and Technology (CONACyT/México) and Enhancement Program for Professors (PROMEP/México).

Esteban Tlelo-Cuautle received a B.Sc. degree from Instituto Tecnológico de Puebla (ITP), México in 1993. He then received both M.Sc. and Ph.D. degrees from Instituto Nacional de Astrofísica, Óptica y Electrónica (INAOE), México, in 1995 and 2000, respectively. From 1995-2000, he was a part of the electronics engineering department at ITP, and in 2001 he was appointed as full professor-researcher at INAOE. From 2009-2010, he served as a Visiting Researcher in the department of electrical engineering at the University of California Riverside. He has authored and co-authored four books, ten book chapters, 45 journal articles and around 100 conference papers. He is an IEEE Senior Member, IEICE Member, and a member of the National System for Researchers (SNI-CONACyT in México). He regularly serves as a reviewer in 18 high impact-factor journals in engineering and 15 internationally recognized conferences. He has been a member of Program Committees in prestigious international conferences. His research interests include systematic synthesis and behavioral modeling and simulation of linear and nonlinear circuits and systems, chaotic oscillators, symbolic analysis, multi-objective evolutionary algorithms, and analog/RF and mixed-signal design automation tools.

PREFACE

This book introduces novel concepts of various areas of physics, computer and electrical engineering. For instance, we present a synthesis methodology to design multi-scrolls chaotic oscillators by dividing the process into three hierarchical levels of abstraction. The first one is the electronic system level, supported by performing high-level behavioural modelling and simulation, and allows exploring a wide range of parameters for the generation of multi-scrolls chaos generators. At this level, to speed-up time simulation, we implement a procedure for the automatic control and determination of step-size in multi-step algorithms. The second level is devoted to use behavioural models of the electronics devices within Verilog-A to describe the hardware of the nonlinear dynamical system. Computer and electrical experts coexist at this level because we show the synthesis of the mathematical descriptions by a block diagram which takes into account the real behaviour of electronic devices. The excursion levels of the chaotic signals are scaled within values that can be synthesized using behavioural models to control the breaking points and slopes of saturated functions. The third level of the Electronic Design Automation (EDA) approach is the selection of an electronic device to implement the chaotic generator. A good repertoire of multi-scrolls chaos generators is presented by dealing with one-dimensional (1D), 2D and 3D oscillators, which respectively requires one, two or three piecewise-linear (PWL) functions. We show the design of the PWL functions by using operational amplifiers (opamps). However, since the high-level approach uses behavioural models, other novel analog integrated circuits (ICs) can be used to implement the chaos generators. To highlight the usefulness of the designed multi-scrolls chaotic oscillators, we present the realization of a secure communication system. Although this application uses commercially available opamps, this is an important part of the book since it opens new lines for future EDA and IC design research, while providing a tutorial for educational purposes.

CHAPTER 1

Introduction

Abstract: In the twenty century advances on the realization of chaotic systems by using electronic devices were introduced. This chapter includes an historical overview of electronics and the realization of chaotic oscillators. From the introduction of the computer, electronic design automation (EDA) born several decades ago, and its usefulness in the synthesis of chaotic systems is described herein. Furthermore, some key-points in performing EDA in dynamical systems at different levels of hierarchy are summarized.

Keywords: Chaos, electronic design automation, dynamical systems, chaos generators, modelling and simulation.

HISTORICAL OVERVIEW

In the beginning of the twentieth century, the flow of electrons could be controlled and electronic circuits were born with the invention of the vacuum tubes. After World War II, the vacuum tubes were replaced by semiconductor devices due to the development of the first bipolar transistor in 1947 [1]. Another milestone was achieved in 1958 with the creation of the first monolithic Integrated Circuit (IC) by Jack Kilby. In 1963, the metal-oxide-semiconductor field-effect transistor (MOSFET) became the basic device for the development of basic building blocks in designing complementary-MOSFETs (CMOS) logic circuits [2]. Thereby, the introduction of ICs opened new possibilities to improve speed, size and reliability [1-3].

Nowadays, designing complex ICs by hand and by using standard CMOS technology became infeasible [4]. Fortunately, some researchers introduced Computer-Aided Design (CAD) methodologies to cope with this problem. For instance, in 1967 the first real application of CAD was reported: a computer program was used to determine the connections between transistors in an IC depending on the specification [1]. At the beginning of the 1980s, a new electronic industry started with the foundation of companies like Mentor Graphics, specialized in tools supporting the IC design process. Today, their Electronic Design Automation (EDA) programs have become indispensable aids for the development of semiconductor products [5-8].

CHAOTIC SYSTEMS

Chaotic systems have been known for long time but only recently, it was shown that chaos could be controlled and therefore, synchronized [9-18]. For this reason; this class of systems promise to have a major impact on many novel, time-and energy-critical applications, such as high-performance circuits and devices (e.g. delta-sigma modulators and power converters), liquid mixing, chemical reactions, biological systems (e.g. in the human brain, heart and perceptual process), crisis management (e.g. in power electronics), secure information processing, and critical decision-making in political, economic and military events [19-22]. This new and challenging research and development area has become a scientific interdisciplinary, involving system and control engineers, theoretical and experimental physicist, applied mathematicians, physiologists and above all, circuit and devices specialists.

Chaos refers to one type of complex dynamical behaviours that possess some very special features such as extreme sensitivity to tiny variations of initial conditions, bounded trajectories in the phase space but with a positive Lyapunov exponent, a finite Kolmogorov-Sinai entropy, a continuous power spectrum, and/or fractional topological dimension, etc [23-27]. In other words, chaos is simply an unpredictable behaviour of a deterministic system in long-term. According to [24], there are three basic identifiers of chaos, namely:

- Chaos is a "deterministic system", i.e. its current condition is a consequence of previous states of the system.

- Chaos is a system that exhibits behaviour which is "difficult to distinguish from random behaviour". In other words, a system may be chaotic if it seems random. This does not mean that the system is random, but rather that the system is difficult if not impossible to predict.

- Chaos is a system that is "sensitive to initial conditions".

Due to these characteristics are nonlinear, their behaviour is much more complicated than that of linear systems. In fact, even the simplest chaotic systems exhibit a bulk of different behaviours [28-31], that can only be fully analyzed with the help of powerful software resources [32, 33-35].

The goal of research for chaotic systems is, thus, to understand how a deterministic dynamical system might exhibit chaotic behaviour, the kind of systems capable of this behaviour, the ways available to control it, the ways to implement it with electronic devices, and the practical and theoretical implications that follow [19,20,36].

Recently, theoretical design and hardware implementation of different kinds of chaotic oscillators have attracted increasing attention [20, 37-41], targeting real-world applications of many chaos-based technologies and information systems [42-58]. It stimulates the current research interest in creating various complex multi-scroll chaotic attractors by using simple electronic circuits and devices [20,59]. Several methods to generate chaotic attractors by using piecewise-linear (PWL) functions, cellular neural networks, nonlinear modulating functions, circuit component design, switching manifolds, etc. have been proposed [60-87]. However, although all these multi-scrolls chaotic attractors have been verified by numerical simulations or theoretical proofs [33-35]; it has been identified that it is difficult to design multi-scrolls chaotic attractors by using electronic devices [17,18,20]. As is well known, it is much more difficult to design a nonlinear resistor that has an appropriate characteristic with many segments [37]. Moreover, the realization of a nonlinear resistor with multi-segments is the basis for hardware implementation of chaotic attractors with multidirectional orientation and with a large number of scrolls [38]. Thereby, circuit implementation of many scrolls needs a larger dynamic range, requiring higher voltage supply and appropriate differential amplifiers or a convenient scaling of voltages [37-39]. However, physical conditions always limit or even prohibit such circuit realization [17,18,20].

Although some approaches exist to design multi-scrolls chaotic systems using electronic devices, those are custom designs [65,70,73,77]. This indicates that it is necessary a deeper understanding on chaotic dynamics of the system [23-27]. In addition, these designs are only valid for a certain range of values for the parameters of the circuit. If one needs to modify the behaviour of the chaotic system, it should be necessary to evaluate the entire design [20].

Considering the aforementioned difficulties, a systematic methodology for circuit design automation of multidirectional multi-scroll chaotic systems must be proposed. The idea is to propose a synthesis approach integrated in an EDA-tool for automatically generating a chaotic system using electronic devices whose characteristics can be adapted by using behavioural modelling and simulation.

APPLICATIONS OF CHAOTIC OSCILLATORS

Chaotic systems have been known for long time, trying to export concepts from physics and mathematics into real-world engineering applications; only recently, it was shown that chaos could be controlled and synchronized between two identical chaotic systems as reported in [9-18]. Chaos has been then considered for use in communications systems [21,22,31]. This was motivated by the extreme sensitivity of chaos to initial conditions and parameters as well as their random like-properties [24]. These properties have emerged in several applications in the communication field [42-58]. Chaos-based communications systems are now in a state of development and possible schemes have been identified and characterized [27,31]. While in the early days of research on chaotic communication, arguments that only some chaos-based circuits and concepts may have relevancy to communication systems were perhaps justifiable, chaotic communication has matured in the last decade to the point where these simple arguments are no longer warranted [27]. The potential applications in communications field are described below.

- **Chaos-Based Modulation Techniques:** Several chaos-based modulation techniques have been proposed in the literature. In [21] is reviewed the state-of-the-art of modulation techniques exploiting chaos. For example in [43] is presented a generalization of frequency modulation in which state trajectories of dynamical systems are used as carrier waves.

- **Coherent Chaos-Based Communication Systems:** The possibility of exploiting the self-synchronizing properties of chaotic systems for communication purposes has attracted the attention of several international researchers over the last decade [9]. For example, in [44] an analytical derivation of bit error rates for multi-user coherent chaos-shift-keying (CSK) communication systems is presented.

- **Noncoherent Chaos-Based Communication Systems:** Noncoherent chaotic communications have evolved from elementary schemes such as COOK (chaotic on–off keying) and CSK (chaos-shift keying) to the more sophisticated differential coherent detection of DCSK (differential CSK) and to the modern FM-DCSK (frequency modulation DCSK) [45].

- **Chaotic Pulse-Position Modulation:** In the last few years, there has been a rapidly growing interest toward ultrawide bandwidth (UWB) impulse radio (IR) communication systems [46,47]. In [47] is shown a UWB system which utilizes chaotically pulse position modulation (CPPM) based on maps.

- **Spread-Spectrum Communications Using Chaos:** In recent years, chaos theory has been applied to the design of spread-spectrum sequences to enhance the performance of DS-CDMA (direct sequence code-division multiple access) systems [48,49]. As shown in [49] where is introduced a DS-CDMA system using chaotic spreading sequences.

- **Filtering of Chaotic Signals:** Noise filtering of chaotic signals exploiting their determinism has received considerable interest in the context of communication systems [50,51]. For example in [51], it is proposed an unscented Kalman filter (UKF) for filtering noisy chaotic signals.

- **Optical Communications Exploiting Chaos:** Recently, researchers from different groups around the world have been trying to develop optical communication systems exploiting chaos [52,53]. For example, [53] presents a study both numerically and experimentally the dynamics, synchronization, and message encoding/decoding for optically injected single-mode semiconductor lasers and single-mode semiconductor lasers with delayed optoelectronic feedback.

- **Data encryption using chaos:** The highly unpredictable and random-look nature of chaotic signals is the most attractive feature of deterministic chaotic systems that may lead to data encryption schemes [11,42,88]. In [11] is introduced the synchronization of multi-scrolls chaotic attractors with applications to private communications. It is demonstrated that the chaotic system is rich in significance and in implication because of sensitivity to change initial conditions, control parameters, random-like behaviour and very high diffusion and confusion properties that are desirable for data encryption.

The future applicability of chaos-based communication systems depends strongly on the development and hardware implementation of reliable nonlinear circuits for generating and processing the chaotic signals [19]. In particular, it is essential to develop efficient and well controllable chaotic generators using simple electronic devices [20,31].

HIGH-LEVEL SYNTHESIS OF ANALOG SYSTEMS

Synthesis is the process of creating a low-level representation (low-level abstraction) from a higher-level (more abstract) representation [1]. The synthesized representation should have the same function as the higher-level representation [89]. Synthesis literally means combining constituent entities to form a whole unit. In circuit design, synthesis refers to the process of finding a set of elements and a way of combining them, such that when so combined to form a circuit, the circuit meets its requirements [7,8]. Whereas synthesis methods for basic building blocks look for optimal values for the parameters, like devices sizes and bias values, to realize a certain set of performance specifications, high-level synthesis or synthesis based on behavioural modelling deals with the generation of the architecture and the selection of its building blocks and their performance values [89-91]. This synthesis allows a design engineer to make decisions at an early stage of the design cycle, thus ensuring correct design [92]. High-level synthesis has become popular because of several advantages it provides [89-91], as discussed below.

- **Continuous and reliable design flow**: The high-level synthesis process provides a continuous and reliable flow from system-level abstraction to basic building block abstraction automatically without manual handling.

- **Shorter design cycle**: If more of the design process is automated, faster products can be made available at lower times. The shorter design cycle can reduce the number of human designers used to design an IC.

- **Fewer errors**: Since the synthesis process can be verified easily, the chances of errors will be smaller. Correct design decisions at the higher levels of circuit abstraction can ensure that the errors are not propagated to the lower levels.

- **Easy and flexible to search the design space**: Because a synthesis system is based on behavioural modelling, the designer has more flexibility to explore the design space and observe the corresponding effects on and the relationships between the performance parameters.

- **Shared knowledge**: As design expertise is moved into synthesis systems, it becomes easier for a non-expert designer to produce a chip that meets a given set of specifications.

This class of synthesis approach enhances the possibilities to create designs of analog systems [5,6]. In general, a behavioural description of a circuit is a representation of the functionality of the circuit by means of an abstract model [90]. The abstraction level is an indication of the degree of detail specified about how a function is to be implemented. Moreover, the abstraction level forms a hierarchy [92]. So far, the lowest level of abstraction is achieved by using device-level descriptions of each of the circuit devices (like, for instance, SPICE-like transistor models). In contrast, behavioural models try to capture as much circuit functionality as possible with far less implementation details than the device-level description of the circuit [93]. Concepts such as *macro*, *high-level*, or *functional* have been commonly used as equivalent adjectives of *behavioural modelling* [91]. Fig. **1** classifies and clarifies the purposes of analog behavioural modelling [5,6, 89-91]; and this classification is explained below.

- **Dimensionality**: This class distinguishes between behavioural models that represent the behaviour of a single circuit (block) [94], from models that represent a range of circuits (system). Indeed, system behavioural models can cover a wide range of electrical characteristics of one or more circuits. Therefore, they are suitable for design exploration and synthesis.

- **Use**: Behavioural models can be built either to carry out both synthesis and exploration in top-down design flows or to be used for verification in bottom-up design flows.

- **Style**: This refers to the nature of the behavioural description of the circuit. A distinction is made between models that represent the behaviour of the circuit by means of an explicit physical model and mathematical models intended to fit some of the responses of the circuit without any physical model. The physical style uses device-level description languages (e.g., SPICE-like syntax) [93]. High-level behavioural models use more abstract description languages, such as the Hardware Description Languages (HDLs) like Verilog-A [91,95], and Verilog-AMS [90].

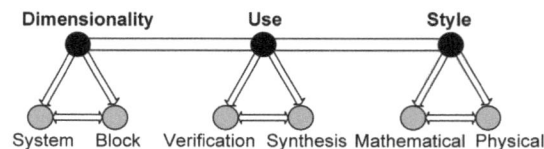

Figure 1: Classification of analog behavioural modelling.

To take full advantage of behavioural modelling, it is necessary to properly define the level of abstraction used in high-level synthesis approaches [89]. For this, a modelling and synthesis strategy must be drawn up in an EDA tool [7]. The former, determines how does an analog system is represented. The description level used is a first criterion to select a modelling style. Furthermore, a good modelling strategy makes it easy to execute the design flow [92] and to support the selected synthesis strategy. The latter, indicates the kind and order of the operations to be applied during the different design stages. It determines issues like how to select a topology, how to change parameters and which details should be added in a refinement step [90-93]. Together, the modelling and synthesis strategies form a design strategy [5]. The adoption of a systematic design methodology results in the identification of design tasks from which EDA tools can be developed.

ELECTRONIC DESIGN AUTOMATION

In electronics, analog circuit design typically needs to take into account several performance specifications, which depend on the circuit designer's abilities to successfully exploit a range of nonlinear behaviours across levels of abstraction from devices to circuits and systems [2-3]. Managing all these aspects is what makes analog circuit design cumbersome and challenging [4-6]. Therefore, efficient synthesis methodologies supported by appropriate analog CAD programs have to be developed.

The IC design effort has been continuously increasing with the integration of more and more functionalities onto a single chip [89]. In the area of CMOS analog design, increasing design complexity is widening the gap between the complex systems and the ability to design them [4-6], [89-92]. From this point of view, in this book, the design complexity for systems is measured according to their dynamics and behaviour [92,93] as described below:

- **Linear Time-Invariant (LTI) Systems** [96]: Any block composed of resistors, capacitors, inductors, linear controlled sources, and distributed interconnect models is LTI (often referred to simply as linear). The definitive property of any LTI system is that the input and output are related by convolution with an impulse response in the time domain.

- **Linear Time-Varying (LTV) systems** [96]: A class of nonlinear circuits (including mixers, switching-filters, and sampling circuits) can be usually described as LTV systems. The key difference between LTV systems and LTI ones is that if the input to an LTV system is time shifted, it does not necessarily result in the same time shift of the output. The system remains linear, in the sense that if the input is scaled, the output scales similarly.

- **Nonlinear (non-oscillatory)** [6,92]: Nonlinearity is, in fact, a fundamental feature of any block that provides signal gain or performs any function more complex than linear filtering. Examples include digital gates, switches, comparators, and analog blocks such as sampling circuits, switching mixers, analog-to-digital converters, etc.

- **Nonlinear (oscillatory)** [27,92]: Oscillators are ubiquitous in electronic systems (including phase-look loops (PLLs)). They generate periodic signals, typically sinusoidal or square-like waveforms that are used for a variety of signal processing applications. From the standpoint of both simulation and modelling, oscillators present special challenges [6]. Traditional circuit simulators such as SPICE consume significant computer time to simulate the transient behaviour of oscillators [90,91]. Furthermore, chaotic oscillators [17,18] that exhibit complex dynamics and strong nonlinear behaviour are becoming an interesting field for research due to its extreme sensitivity to initial conditions as well as their random-like properties [20].

The key to deal with this diversity and dynamics of nonlinear systems in general lies in adopting appropriate and well-structured synthesis methodologies, which must be supported by efficient EDA tools [5-7]. Therefore, the goal of analog design automation (ADA) is to reduce the manual design time required for designing an analog circuit [4,7] with the help of a CAD resource that allows human designers built their designs while satisfying the performance specifications [8].

Whereas the first automation efforts were limited to specific tasks CAD programs, they are now encountered at various stages throughout the design flow [89]. Some examples of application areas are simulation, physical design, analysis, design-for-test, verification and synthesis [89]. Analog CAD tools for low-level tasks like generation of transistor layouts, extraction of parasitic elements and verification via simulations are now well established [97]. Tools of a second generation provide solutions for analysis via numerical and symbolic derivation of performance characteristics [98], and for the selection of values for parameters like transistor sizes and capacitor values in basic building blocks [99]. This book, however, focuses on a third generation of analog CAD tools [100], which look to a higher level of abstraction covering nonlinear analog systems [89-91], and to corresponding simulation, modelling and synthesis methodologies [92].

PURPOSE OF THIS BOOK

The aim of this book is to present contributions to the ADA of nonlinear dynamical systems, specifically, multi-scrolls chaotic oscillators based on saturated nonlinear function series, with multi-segments. The problem, and, hence, the goal of the research here reported, consists in the improvement of the analog design process for chaotic systems by means of the definition, development, and demonstration of an automatic synthesis methodology, which, attending to the issues mentioned above, helps to design this class of systems and allows to exploit their possibilities in future applications (e.g. chaos-based transmission systems [31]). The purpose of this research can be summarized around the main issues of the ADA of chaotic systems, corresponding to the main parts of this book.

- **Multi-scrolls chaotic oscillators design**: First, from the previous sections, it was identified that implementation of reliable nonlinear dynamical circuits for generating various complex chaotic signals is a key issue for future applications of chaos-based transmission systems. In particular, creating various complex multi-scrolls chaotic attractors by using some simple electronic devices is a topic of both theoretical and practical interests [101]. However, it is difficult to design multi-scrolls chaotic systems by an analog circuit because it exhibits complex dynamics and strong nonlinear behaviour, extreme sensitivity of chaos to initial conditions as well as their random-like properties. Therefore, this book introduces an ADA methodology focused on multi-scrolls chaotic systems. This strategy consists of a top-down hierarchical design flow based on behavioural modelling integrated in a synthesis methodology.

- **Behavioural modelling**: A modelling strategy suited for analog synthesis should deal with the ability to represent the chaotic behaviour for different control parameters of chaotic systems. This requires using a behavioural model that is generic enough to facilitate the representation of various chaotic systems. In particular, the multi-scrolls chaotic systems are modelled at the highest level of abstraction, i.e. Electronic System Level (ESL), by applying state variables approach and PWL approximations. Further, time-efficient evaluation of chaotic behaviour, via numerical simulation, is necessary for extensive design space exploration. Therefore, a modification for the third-order Adams-Moulton algorithm is given. Two main procedures to modify characteristics of multi-scrolls chaotic oscillators are introduced: excursion levels scaling and frequency scaling.

- **Automatic synthesis**: The translation between high-level descriptions and lower levels in the top-down ADA flow for analog systems that exhibit complex dynamics and strong nonlinear behaviour, such as chaotic oscillators, is a challenging issue. Therefore, this book introduces a solution to that problem by proposing a new synthesis methodology for designing multi-scrolls chaotic systems by using behavioural modelling; where the system-level specifications are transmitted down to obtain performance specifications for every circuit-level component, e.g. operational amplifiers (opamps), of the chaotic system. In this book, this process is carried out by using a basic cell based on opamps. The opamp finite gain model is related to breakpoints and slopes of the PWL functions (nonlinear functions with multi-segments are needed to generate multi-scrolls chaotic attractors). To speed-up time simulation, a Verilog-A model for opamps is used at this level.

- **CAD-tool prototype**: This book introduces guidelines to implement a CAD-tool prototype for the electronic design automation (EDA) of multi-scrolls chaotic dynamical systems by including the three previous points.

In the second chapter, the backgrounds of the chaotic systems and behavioural modelling are summarized. The behavioural modelling strategy for the multi-scrolls chaotic systems by applying state variables approach and PWL approximation is further elaborated in chapter three. Two novel procedures are defined in this chapter. The first one uses an extra parameter to control the slopes and breakpoints of the PWL function and in this manner; the excursion level scaling of chaotic signals is executed. For the second procedure, a frequency-scaling framework has been developed which allows choosing the frequency for the chaotic attractors. Besides, a new scheme to automatically control and determine the step-size for the third-order Adams-Moulton algorithm is also presented. Last but not least, a procedure to calculate the Lyapunov exponents of the multi-scrolls chaotic systems is described.

In the fourth chapter, the ADA process is systematically analyzed and an approach for the automatic synthesis of saturated nonlinear function series is introduced. By comparing the properties of the saturated nonlinear function series with the parameters of the opamp finite gain model, one obtains a novel opamp-based basic cell to synthesize saturated functions with multi-segments. This design approach involves a modelling strategy using Verilog-A behavioural models or macromodels for the opamps leading to SPICE simulations to explore a design space containing different models for the electronic devices.

In the fifth chapter, the proposed synthesis methodology to design multi-scrolls chaotic oscillators is presented. In particular, it is shown the synthesis of chaotic systems using opamps. The resulting attractors are multiple orientations on phase space: one-dimensional orientation (1D), two-dimensional orientation (2D) and three-dimensional orientation (3D). SPICE simulations results show the usefulness of the proposed synthesis methodology. This top-down approach naturally deals with diverse behaviour characteristics and explores the design space. Details on the implementation of an EDA environment to design multi-scrolls chaos generators are described. Particular attention is giving on the generation of chaotic attractors on 1D, 2D, 3D-orientation for different sets of high-level specifications. Besides, a secure communication scheme based on Hamiltonian forms approach to synchronize the chaotic systems and chaotic additive masking to transmit analog signals is described. In this manner, experimental realizations confirm the theoretical proofs and the SPICE simulations results.

CHAPTER 2

General Theory

Abstract: The preceding chapter introduced the motivation for further improving in automated synthesis methodologies for chaotic systems. Increasing the efficiency of synthesis, however, can be achieved if one uses higher abstraction levels to capture the behaviour of the analog systems. Behavioural modelling is seen as a promising solution that, combined with the appropriate synthesis tools, can be used to design multi-scrolls chaotic systems. Hence, this chapter first describes the basic concepts about nonlinear dynamical systems and the existing approaches to design chaotic systems as well as the applications of this class of systems. Afterwards, several important aspects in behavioural modelling are introduced. The steps performed during different design stages are represented by identifying the abstraction and description levels. Finally, the approaches for EDA of nonlinear dynamical systems are reviewed, and the interaction of four basic operations is discussed to introduce the definition of a synthesis methodology for analog designs based on behavioural modelling.

Keywords: Chaos, electronic design automation, dynamical systems, chaos generators, modelling and simulation.

CHAOTIC SYSTEMS

Chaotic systems refer to one type of complex nonlinear dynamical system that possesses some very special features such as extreme sensitivity to tiny variations of initial conditions and parameters, and bounded trajectories in the phase space but with a positive maximum Lyapunov exponent [23-27]. Lately, chaotic systems promise to have a major impact on many novel applications [42-58].

For deterministic chaos to exist, a nonlinear dynamical system must have a dense set of periodic orbits, be transitive, and sensitive to initial conditions [23]. Density in periodic orbits implies that any periodic trajectory of the orbit visits an arbitrarily small neighborhood of a non-periodic one [24]. Transitivity relates to the existence of points a, b for which a third point c can be found that is arbitrarily close to a and whose orbit passes arbitrarily close to b [24]. Finally, sensitivity to initial conditions is the property to arbitrarily close initial conditions to give rise to orbits that are eventually separated by a finite amount [25]. In the following section, basic concepts on dynamical systems are given, followed by a review of approaches in the analog circuit design of chaotic systems as well as some applications.

BASICS OF DYNAMICAL SYSTEMS

A dynamical system is characterized by a set of related variables, which can change with time in a manner which is, at least in principle, predictable provided that the external influences acting on the system are known [26]. This book deals only with deterministic systems, that is to say, it takes no account of statistical properties [27].

State-space Models

In order to make any quantitative progress in understanding the behaviour of a system, a mathematical model is required. Such a model may be formulated in many ways, but their essential feature is to enable us to predict the future behaviour of the system, given its initial condition and knowledge of the external forces which affect it [23]. The mathematical structure most naturally adapted for this purpose is the so-called state-space representation [23,102], which consists of a set of differential equations, describing the evolution of the variables whose values at any given instant determine the current state of the system. These are known as the state variables and their values at any particular time are supposed to contain sufficient information for the future evolution of the system to be predicted, given that the external influences (or input variables) which act upon it are known. The differential equations must, therefore, be of first order in the time-derivative, so that the initial values of the variables will suffice to determine the solution. For convenience of notation, it is common to collect the state variables into a vector $\mathbf{x}$ (the state vector), the input variables into a vector $\mathbf{u}$ (the input vector), and the output vector $\mathbf{y}$ and write the equations in the form [25]:

$$\dot{\mathbf{x}} = \mathbf{f}(\mathbf{x}, \mathbf{u}, t)$$
$$\mathbf{y} = \mathbf{h}(\mathbf{x}, \mathbf{u}, t)$$

(1)

where the dot denotes differentiation with respect to time (t), and the functions $\mathbf{f}$ and $\mathbf{h}$ are, in general, nonlinear. Nonlinear functions may arise in a dynamical model either because they are intrinsic to the nature of the system or because, in a technological case such as a control system, they have been deliberately introduced by the designer for a specific purpose [27]. The variety of possible nonlinearities is infinite, but it may nevertheless be worthwhile to classify them into some general categories [26].

First, there are simple analytic functions such as powers, sinusoids and exponentials of a single variable, or products of different variables. A significant feature of these functions is that they are smooth enough to possess convergent Taylor expansions at all points and thus, can be linearized [23]. A type of nonlinear function frequently used in system modelling is the PWL approximation [103,104], which consists of a set of linear relations valid in different regions. They have the advantage of changing from nonlinear dynamical to linear equations (and hence solvable) in any particular region, and the solutions for different regions can then be joined together at the boundaries.

Although the target is mainly concerned with nonlinear phenomena, it is appropriate at this point to review the special case of linear systems, partly because linear approximations are so widely applicable to solve nonlinear systems [27,96]. For systems with finite dimensional state-space representations, the equations describing a linear model become:

$$\dot{\mathbf{x}} = \mathbf{A}\mathbf{x} + \mathbf{B}\mathbf{u}$$
$$\mathbf{y} = \mathbf{C}\mathbf{x} + \mathbf{D}\mathbf{u}$$

(2)

where $\mathbf{A}$, $\mathbf{B}$, $\mathbf{C}$ and $\mathbf{D}$ are matrices (possibly time-dependent) of appropriate dimensions. The great advantage of linearity is that, even in the time dependent case, a formal solution can immediately be constructed, which is moreover applicable for all initial conditions and all input functions [24].

However, an important point which must be kept in mind for a nonlinear dynamical system is that the stability properties are essentially more complicated than in the linear case, and in particular, it is necessary to distinguish between local and global aspects [23]. For a linear system, there is no such distinction, but when nonlinearities are present, several new features can appear such as limit cycles or the phenomenon known as *chaos* [23-27]. In any case, the type of behaviour actually manifested by a nonlinear dynamical system, whether stable, unstable, oscillatory or chaotic, may depend critically on the input applied to it. Thus differs from the linear case where all the dynamical properties can be described, for example by a transfer function, independently of the input.

Autonomous Systems

Although the equations of a dynamical model will, in general, depend on the time, either explicitly, through the input function, or both, a large part of nonlinear system theory is concerned with cases where there is no time dependence at all [25]. Such systems are said to be autonomous, and they arise quite naturally in practice when, for example, the input vector is held fixed [23]. In any such case, the differential equation for the state vector will become:

$$\dot{\mathbf{x}} = \mathbf{f}(\mathbf{x}, \hat{\mathbf{u}})$$

(3)

where $\hat{\mathbf{u}}$ is a constant vector. Thus, the equilibrium points in the state-space are determined by $\mathbf{f}(\mathbf{x}, \hat{\mathbf{u}}) = 0$. Assuming that $\mathbf{f}(\mathbf{x}, \hat{\mathbf{u}})$ satisfies Lipschitz condition [102], the differential equation for $x(t)$ will have a unique solution, for any given initial state $x(0)$. The path traced out in the state-space by $x(t)$ is called a trajectory of the system and because of the uniqueness property [33], there will be one and only one trajectory passing through any given point. If it is suppressed the dependence on $\hat{\mathbf{u}}$, the state-space differential equations for an autonomous system can be written simply as:

$$\dot{\mathbf{x}} = \mathbf{f}(\mathbf{x})$$

(4)

and the set of all trajectories of this equation provides a complete geometrical representation of the nonlinear dynamical behaviour of the system, under the specified conditions. As a result, it is possible to give an essentially complete classification of behaviour in the phase plane, though not in higher-dimensional state-spaces [35]. In general, the equations describing a nonlinear dynamical system cannot be solved analytically, so that, in order to construct the trajectories accurately, it is necessary to use numerical methods [34].

Equilibrium Points

The equilibrium points of an autonomous system given by (4) are also known as singular points when $\mathbf{f}(\dot{\mathbf{x}}) = 0$; because they appear to violate the general rule that only one trajectory can pass through any given point [27]. Actually, the violation is only apparent, since the trajectories which meet at a singular point do not really pass through it, but only approach or depart from it asymptotically [24]. Assuming that $\mathbf{f}(\mathbf{x})$ is smooth enough for the equations to be linearized around the singular point $\hat{\mathbf{x}}$, the approximation will be sufficient to determine the behaviour of the trajectories in the neighborhood of the equilibrium point. If $\mathbf{A}$ in (2) is nonsingular, the nature of the equilibrium is essentially determined by its eigenvalues and can be classified by stable node, stable focus, unstable node, unstable focus, centre and saddle point [23-27], as illustrated in Fig. **1**.

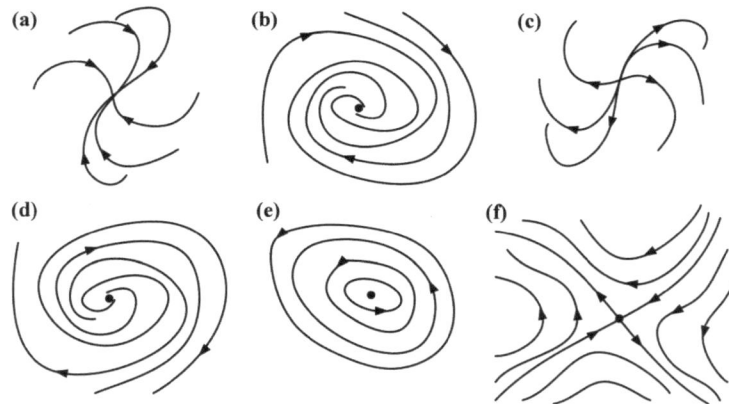

Figure 1: (a) stable node, (b) stable focus, (c) unstable node, (d) unstable focus, (e) centre and (f) saddle point.

Limit Cycles

A common feature of autonomous systems is the occurrence of a special type of trajectory which takes the form of a closed curve. This is known as a limit cycle and represents a periodic solution of the system equations since, when the state vector returns to its initial value, it must repeat its previous motion and so continues indefinitely [26]. Limit cycles can occur in systems of any order, and indeed constitute the typical form of oscillatory behaviour which arises when an equilibrium point of a nonlinear system becomes unstable according to the general condition for the existence of limit cycles defined by the Poincaré-Bendixson theorem [26]. Another concept also due to Poincaré, which is relevant to the occurrence of limit cycles in the phase plane, is the *index* of a closed curve [23]. If the curve is simple, that is to say, it does not intersect itself, its index with respect to the vector function $\mathbf{f}(\mathbf{x})$ is defined as the net total number of clockwise revolutions made by $\mathbf{f}$ as $\mathbf{x}$ traverses the curve once in the clockwise sense. Furthermore, it implies that the index of the curve can be computed by summing the contributions of the singular points which it surrounds, assuming they are isolated, where each node, focus, or centre counts +1, and each saddle point counts -1 [33]. Since the index of a limit cycle is clearly +1, this restricts its possible location with respect to the equilibrium points of the system. A typical case [27], with a stable limit cycle surrounding an unstable focus, is illustrated in Fig. **2**.

Strange Attractors and Chaos

Although singular points and closed curves constitute the only asymptotic terms of bounded trajectories for autonomous systems in the phase plane, this is no longer true in spaces of higher dimension [24]. In general, the term for a limit set where all trajectories in its vicinity approach it as $t \to \infty$; is an *attractor* [26], since it asymptotically attracts nearby trajectories to itself. For second-order systems, the only types of limit set normally encountered are singular points and limit cycles. Consequently, a continuous-time autonomous system requires more than two dimensions to exhibit chaos [23-27]. Therefore, in a state-space of more than two dimensions, a far greater variety of behaviour is possible; for example, a *torus* [35]. More complicated still are the so-called strange limit sets [23]. There may or may not be asymptotically attractive to neighboring trajectories; and if so, they are known as *strange attractors* [23-27]. Even then, the trajectories they contain may be locally divergent from each other, within the attracting set. Such structures are associated with the quasi-random behaviour of solutions called *chaos* [24].

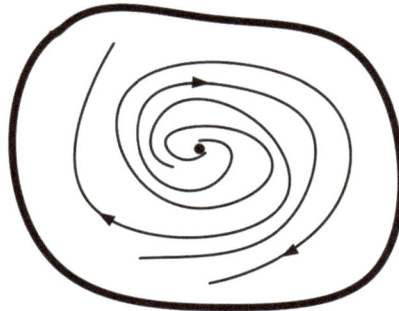

Figure 2: Limit cycle surrounding an unstable focus.

As a corollary, no chaotic dynamical system can have real fixed points within its attractor. In general, the equations that specify a dynamical system are dependent on a parameter or set of parameters and chaotic behaviour only manifests itself for certain values of these parameters [23,26]. It is only under the chaotic regime that the system cannot have real fixed points [23]. For example, Rössler system exhibits chaotic behaviour as shown in Fig. **3** [33].

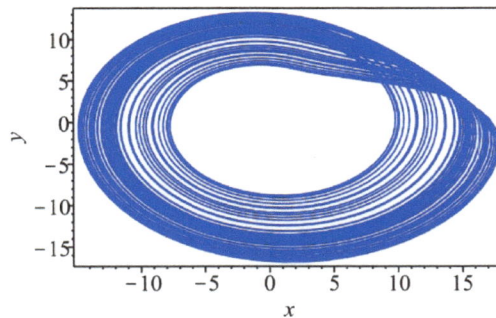

Figure 3: Strange or chaotic attractor for Rössler system.

APPROACHES IN THE EDA OF CHAOTIC OSCILLATORS

Over the last two decades, theoretical design and circuit implementation of various chaos generators have been a focal subject of increasing interest due to their promising applications in various real-world chaos-based technologies and information systems [60-87]. This subsection offers an overview of the subject on multi-scroll chaotic attractors generation, including some design methodologies and circuit implementations. In particular, Chua's circuit [28-31], as a paradigm of chaos and a bridge between electronic circuits and the chaos theory, has been widely studied and used as a platform for engineering applications [105]. Recently, extended from Chua's circuit, theoretical design and hardware implementation of different kinds of chaotic oscillators have attracted increasing attention, especially for those that can create various complex multi-scroll chaotic attractors by using simple electronic circuits and devices [65,70,73,77].

Design of Multi-scroll Chaotic Attractors Via PWL Functions

In this subsection, the multi-breakpoint PWL function [103] approach is introduced, which can generate multi-scroll chaotic attractors from Chua's circuit or other simple nonlinear systems. The main idea there is introducing additional breakpoints in the PWL function [65].

In [60], an approach to design n-double scroll chaotic attractors using quasi-linear functions is introduced. This method is based on Chua's circuit and is a kind of qualitative method, but it gives a global insight into the dynamic behaviour of a dynamical system and it can be viewed as complementary to the conventional linearization approach. Although, this method can generate n-double scroll attractors, those scrolls will be invisible for larger *n* due to that dimension of the scrolls near to origin will become very small. Another approach to design multi-scroll attractors is introduced in [61], where a generalized Chua's circuit is proposed. It is based on a PWL function with multi-segments. This method can also generate n-double scroll attractors. A family of multi-scroll chaotic attractors generated from the unfolded Chua's circuit is introduced in [62].

Regarding the implementation of multi-scroll chaotic attractors, [63] and [64] have experimentally confirmed an n-double scroll chaotic attractor by using a state-controlled cellular neural network (CNN)-based circuit. It was demonstrated that generalized Chua's circuit was equivalent to a single-layer three-cell state-controlled CNN. Another implementation of the generalized Chua's circuit is shown in [65], where 3 and 5-scrolls were experimentally verified by using a voltage-controlled voltage-source (VCVS) implementation. By using the scaling properties of nonlinearity in a generalized Chua's circuit, the authors in [66] experimentally confirmed a 6-scrolls chaotic attractor. Also it is shown that by re-scaling the breakpoints in the PWL function by an appropriate factor, all the slopes remain the same. Lately, [67] proposed a circuit design method for experimentally verifying a maximum of 10-scroll chaotic attractors by introducing additional breakpoints into the PWL function for Chua's circuit. However, it is necessary to remark that it is quite difficult to create attractors with a large number of scrolls due to the limitation of the real dynamic range of the available physical electronic-devices [20]. Finally, [68] proposed an improved method for generating multi-scroll chaotic attractors. A general recursive formula was derived for determining the equilibrium points and breakpoints in voltage, and in this manner, an 11-scroll chaotic attractor was experimentally observed.

Design of Multi-scroll Chaotic Attractors Via Basic Circuits

This subsection introduces several approaches for generating multi-scroll chaotic attractors by using basic circuits. It is well known that the step circuit, hysteresis circuit and saturated circuit are the three types of basic circuits [20,96]. First, a family of scroll and grid-scroll attractors by using the step circuit, including 1-D (one-dimensional) n-scroll, 2-D n × m-grid scroll, and 3-D n × m × l-grid scroll chaotic attractors is introduced in [75]. The state equations of this family of systems depend on the number of nonlinear functions to generate multi-scrolls. This indicates that three nonlinear functions are needed to generate 3D-scrolls. Also, it is designed a circuit diagram using commercial current-feedback operational amplifiers (CFOAs) [106] and comparators. Then, a systematic approach is introduced for generating multidirectional multi-scroll chaotic attractors by using hysteresis series. It includes two cases: the system to be controlled is a 2D linear autonomous system, and a three-dimensional (3D) linear autonomous system. The former is proposed in [76] and its circuit implementation consists of two function parts: (a) a hysteresis series building block; and (b) the second-order system. Both are implemented using opamps [107], and two diodes for the hysteresis block. A 2-D 5×3-grid scroll attractor was experimentally observed. The latter introduces a systematic approach for generating multidirectional multi-scroll chaotic attractors using hysteresis series [77,78].

Finally, [79] introduces a saturated multi-scroll chaotic system based on saturated nonlinear function series. This system can produce three different types of attractors, as follows: 1-D saturated n-scroll chaotic attractors, 2-D saturated n × m-grid scroll chaotic attractors and 3-D saturated n × m × l-grid scroll chaotic attractors. Moreover, in [79] is also constructed a 2D Poincaré return map to rigorously prove the chaotic behaviours of such saturated double-scroll attractors. The mentioned above circuit design method provides a theoretical principle for circuit implementation of such chaotic attractors with multidirectional orientations and a large number of scrolls.

Design of Multi-scroll Chaotic Attractors Via Other Techniques

In this subsection, other design techniques for generating chaotic attractors are reviewed. The first one is based on nonlinear modulating functions such as the sine function [69], the smooth hyperbolic tangent function [70], and the switching signum function [71]. Generalizations of those methods are presented in [72-74], where multi-scrolls chaotic attractors are generated from the general jerk circuit [101]. Another technique is based on switching-manifold control applied to some simple linear systems for generating multi-scrolls chaotic attractors as shown in [80-82].

A modified Lorenz system introduced in [83,84], is free from the positive z constraint. That system is a dual of the original Lorenz and can generate up to four-wing butterfly attractors. Also, in [83] is designed a circuit diagram to experimentally verify the four-wing butterfly chaotic attractor. It mainly includes capacitors, MOS analog switches, CFOAs and a reference voltage. Another approach based in Lorenz system is shown in [85]. Based on the proto-Lorenz system, one can design and generate up to 4-scroll chaotic attractors by introducing a linear transformation in the original Lorenz system.

In [86] is introduced a critical chaotic system, which can display (i) two 1-scroll chaotic attractors simultaneously, with only three equilibriums, and (ii) two 2-scroll chaotic attractors simultaneously, with five equilibriums. That system is found to be chaotic in a wide parameter range and has many interesting complex dynamical behaviours. Finally, one simple fractional order system is constructed for generating multi-scroll chaotic attractors [87]. Notice that for an integer order case, that system can generate double-scroll chaotic attractors and for a fractional order case, the range of parameters for creating chaos depends on the given fractional order.

It should be pointed out that several design methods for generating multi-scroll chaotic attractors have been developed by using piecewise-linear (PWL) functions, cellular neural networks, nonlinear modulating functions, circuit component design, switching manifolds, basic circuits, and so on. However, it has not been developed a synthesis methodology for automating the design process for multi-scroll chaotic attractors [19,20]. From this point of view, this book is focused on proposing guidelines to implement an automatic synthesis methodology to design multi-scrolls chaotic systems by using behavioural modelling [89].

BEHAVIOURAL MODELLING

Engineers analyze and design various types of systems. In general, a system can be defined as a collection of interconnected components that transforms a set of inputs received from its environment to a set of outputs [4]. In an electronic system, the vast majority of the internal signals used as interconnections are electrical signals. Inputs and outputs are also provided as electrical quantities, or converted from, or to, such signals using sensors or actuators [2]. Actually, behavioural modelling can be a possible solution for the successful development of analog EDA tools due to various types of systems that can be represented by means of an abstract model [90,91]. The abstraction levels are indications of the degree of detail specified on how the function is to be implemented [7]. Therefore, behavioural models try to capture as much circuit functionality as possible with far less implementation details than the device-level description of the circuit [5]. The next subsections explain the need for a behavioural view and review its most general characteristics as well as the approaches in analog synthesis.

Description and Abstraction Levels

Systematic synthesis methodologies based on different design representations are commonly used when designing both analog and digital electronic systems [89]. However, it is more difficult to make a strict distinction between different abstraction levels for analog systems in contrast to common practice in digital synthesis methodologies [5]. Instead, a division should be made between a description level and an abstraction level as described below [92].

- **Description level**: A description level is a pair of two sets; a set of elementary elements and a set of interconnection types. Five description levels are commonly used as shown in Table 1.

- **Abstraction level**: The abstraction level of a description is the degree to which information about non-ideal effects or structure is neglected compared to the dominant behaviour of the entire system.

Whereas a description level indicates how the analog system is represented; an abstraction level deals with the relation between the model of the system and its real behaviour [93]. Abstraction levels obtain their meaning from the comparison with other abstraction levels. By making the description more understandable via simplification, the abstraction level is increased. On the other hand, adding more details lowers the abstraction level [8]. When information about the structure of the system is added to the representation, the hierarchy of descriptions is traversed towards a lower abstraction level. A system may be described with a certain description level at different levels of abstraction [5]. Although it is clear to consider the functional level at a 'high' abstraction level and the physical level at a 'low' abstraction level, it is not straightforward to compare the abstraction levels of different description levels

[6]. Fig. **4** indicates which areas in the abstraction-description plane are usually covered during an analog design flow [90]. Due to the overlap one can easily jump, for instance, from the behavioural to the circuit level.

Table 1 Description levels commonly used in the design flow of integrated analog systems.

Physical level description	Circuit level description	Macro level description
The system is described as a collection of planar geometric shapes (patterns of metal, oxide, etc.) corresponding to the physical layout of the IC as shown below.	A connection graph between basic elements (e.g., transistors, resistors, capacitors) represents the system. These basic elements will be translated into the physical description by mapping individual or a small collection of them onto a set of interconnected rectangles. The circuit level description for a one-stage OTA is shown below	Macro-models are used to describe the system. They are built up out of controlled sources, inductances, op amps, switches, etc. No one-to-one relation between the elements of the macro-model and the circuit elements can be identified. A number of elements in the macro-model can represent a single element in the circuit. The macro level description for a BJT transistor is shown below.
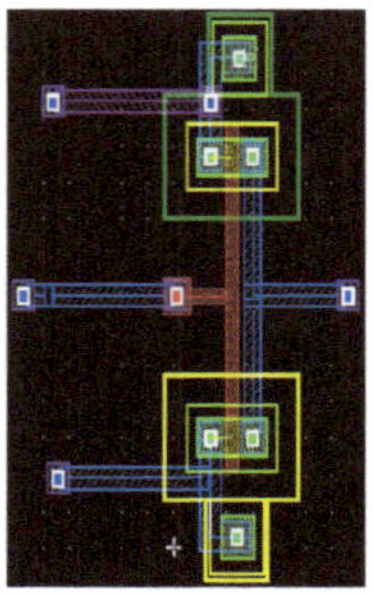	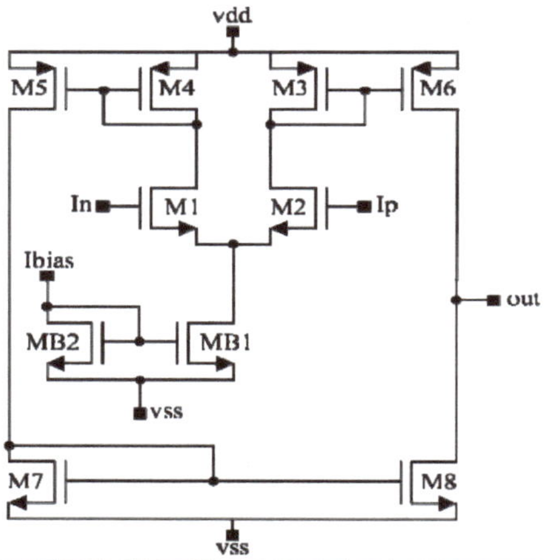	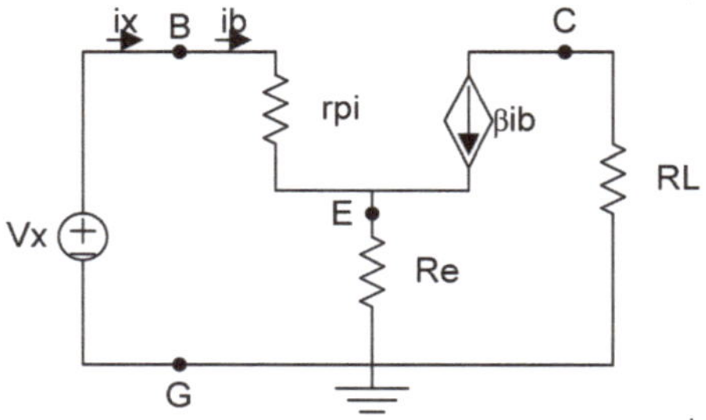

Behavioural level description	Functional level description
The system consists of a collection of building blocks which are described by a set of mathematical relations between the input and output signals found at their ports. The behavioral level description for an analog switch is shown below.	Mathematical equations describe how the input information signals are mapped onto the output information signals. This operation can be represented by a signal flow graph. The functional level description for a biquad filter is shown below.

```
//VerilogA for pipeline
`include "constants.vams"
`include "diciplines.vams"

module switch(p,n,ps,ns);
  input ps, ns;
  parameter real thresh=0,
ron=0.001 from (0:inf);
  parameter real goof=0 from
(0:1/ron);
  electrical p,n,ps,ns;

analog
  begin
    @(cross(V(ps,ns)-thresh,0))
      begin
        if(V(ps,ns)>thresh)
          begin
            V(p,n)<+ I(p,n)*ron;
          end
        else
          begin
            I(p,n)<+ goff*V(p,n);
          end
      end
  end
endmodule
```

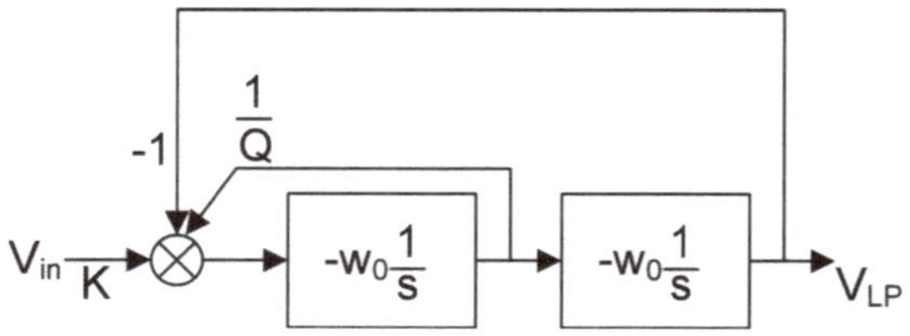

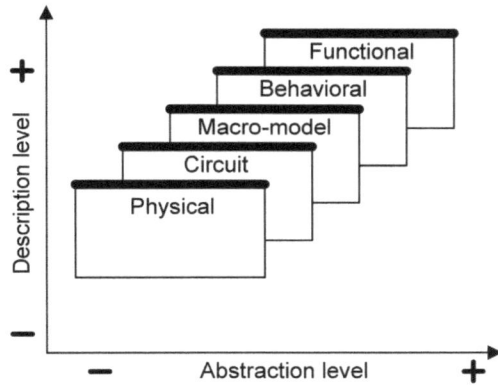

Figure 4: Relationship between description and abstraction levels.

Four Basic Operations

An electronic system is designed by converting the functional specification at the highest abstraction level to a physical realization at the lowest abstraction level via operations between description and abstraction levels. Thereby, automatic synthesis of analog systems can formally be represented as operations between different levels in Fig. **4** as shown in Fig. **5**. Four fundamental types of such operations are distinguished in [90-93]. They are denoted by refinement, simplification, translation and transformation, which are introduced below.

- **Refinement**: Translates a system described with a certain description level, into a representation at a lower abstraction level with the same description level.

- **Simplification**: Translates a system described with a certain description level, into a representation at a higher abstraction level with the same description level.

- **Translation**: Translates a system described with a certain description level, into a representation with another description level preserving the level of abstraction.

- **Transformation**: Translates a system described with a certain description level, into a representation at the same abstraction level and the same description level.

Systematic synthesis implies the application of subsequent *refinement operations* on the system [5]. Such an operation may introduce more detailed information about the system and therefore the performance of the system can then be re-evaluated. A refinement operation can also be used to introduce a subdivision of the system [6]. Hence, additional knowledge about the structure or building blocks of a system corresponds to descending the hierarchy of descriptions. Such operation also includes the derivation of values for the parameters used in the models of the building blocks.

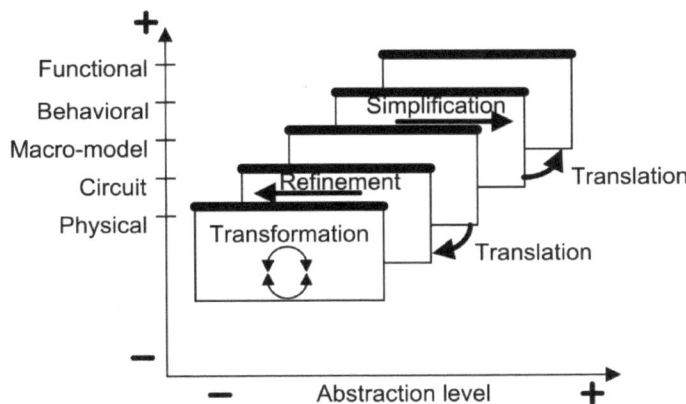

Figure 5: The four basic operations between description and abstraction levels.

Model generation consists of applying multiple *simplification operations* to a system representation to obtain a model with less accuracy, but is easier to interpret or simulate [93]. Simplification is the reverse operation of refinement. A *translation operation* usually happens between two adjacent description levels, although a designer can jump over one level. Consequently, a translation describes an operation in both directions [91]. An example is the mapping of a transfer function into an opamp circuit (from behavioural to macro level). Since a *transformation operation* does not increase or decrease the level of abstraction, applying transformations allows the exploration of different interconnection schemes between the elementary elements [7]. A transformation always consists of two parts. First, a new architecture is generated by choosing another set of elementary elements and its corresponding interconnections. Then, the values for the parameters of the elementary elements should be defined [5,6]. The designer has to deal with the fact that when applying the operations, the behaviour of the system is altered if the abstraction level is changed. An analog system cannot be designed at a certain abstraction level without knowledge of other abstraction levels [90]. These properties are needed to guarantee consistent functionality between systems at different abstraction levels. However it is necessary to correctly define the description and abstraction levels and the operations among them to propose novel synthesis methodologies [92].

DESIGN FLOW FOR ELECTRONIC SYSTEMS

A design process formally consists of the application of a chain of operations defined in the previous section from specification (High-level) to implementation (Low-level) [5,6]. As shown in Fig. **4** the design starts from a description of the functionality of the system, possibly written in some HDL such as Verilog-A, and ends with a layout ready to be fabricated. The four basic operations shown in Fig. **5** can be put together in an iterative design process shown in Fig. **6**, where if the design fails to meet the specifications, a transformation of either the architecture or its parameters should be applied [92]. Simplification can remove details to make it easier to choose which transformation should be selected [90]. On the other hand, if the specifications are met, but the abstraction level does not correspond to the wishes of the designer, details should be added or removed by applying simplification or refinement operations. Finally, the design within the description level finishes once all specifications are met. Translation will be required to convert the current representation of the system to the next description level in either top-down or bottom-up direction [89]. The design flow in Fig **6** is a framework that fits most tasks supported by EDA-tools for analog systems [5-8]. However, selection of a design methodology for a particular case (e.g. chaotic systems) depends on the system itself [5]. The actual tasks to perform during the synthesis process depend on the properties of the systems. During the design of an analog system, the abstraction level can be subsequently lowered or raised, or left unchanged. These three different ways to cope with the abstraction level correspond to the three major design methodologies that are frequently adopted in analog synthesis [7,92].

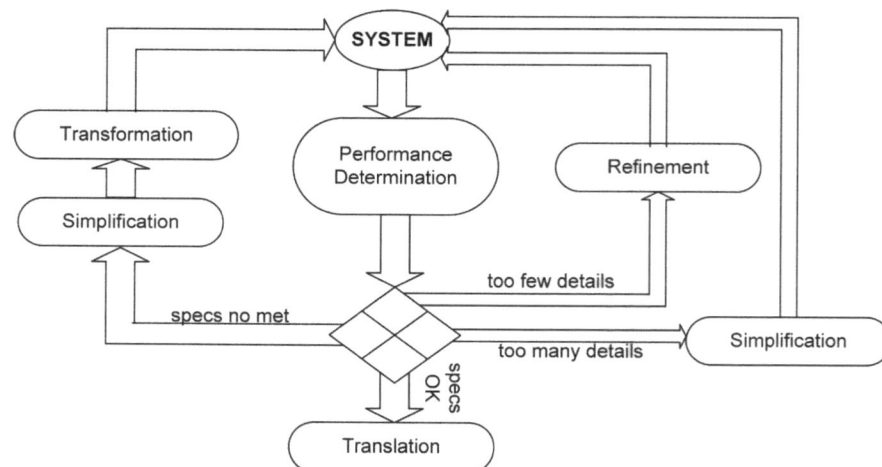

Figure 6: Iterative analog design process using the four basic operations.

Flat Methodology

In this approach, a translation operation is performed usually directly to the circuit level. Neither refinement nor simplification steps are used. Instead, design plans are used to find a set of parameters. The main disadvantage of this methodology is the limited complexity of the designs it can handle [5]. Its main application is the design of basic building blocks like opamps [8].

Bottom-up Methodology

The design process most often starts with a description at circuit level of the building blocks. They are synthesized independently from each other after which the entire system is assembled. Therefore, no refinement operations are needed in the design flow [7]. The major challenge lies in ensuring correct behaviour of the system after assembling and depends on the experience of the designer. It is not applicable for system-level exploration [92].

Top-down Methodology

To cope with complex electronic designs starting from functional specifications, a large system is divided into smaller blocks in the top-down design methodology [90]. The design at a high abstraction level of a complex system corresponds to deriving the behavioural models for the building blocks [89]. Simplification operations are not used in this method.

From this point of view, the EDA methodology must indicate the kind and order of the operations to be applied during the design process and include an appropriate modelling strategy to determine how a system is represented [90,91]. Selecting a good modelling strategy make it easier to execute the synthesis process.

SYNTHESIS APPROACH

According to the definitions introduced in the previous section, the following division for existent synthesis methodologies is obtained [7].

Knowledge-based Synthesis

The basic idea of knowledge-based synthesis is to have a predefined design plan to find and combine the elements such that the requirements (for sizing or layout) are met. The design plans, in the form of design equations, design heuristic strategies, or both, are implemented to represent the steps that an expert designer could take to design a circuit. Then, the underlying principle is to capture the expertise (knowledge) of a designer, so that the synthesis method can obtain an optimum solution using the captured experience. Used for sizing, the design equations are formulated to give the performance characteristics for the system. Examples of knowledge-based sizing tools are IDAC [109], OASYS [110], BLADES [111], CADICS [112] and MIDAS [113]. Used for layout synthesis of IC, the intended captured knowledge refers to the procedures that expert layout designers use to improve the quality of the layout from specific placement strategies to routing techniques. There are two types of knowledge-driven approaches, namely, rule-based and template-based approaches [114,115].

Optimization-based Synthesis

In these tools, the synthesis problem is translated into a function minimization problem that can be solved through numerical methods. For sizing, the basics of optimization-based synthesis are illustrated as being updated at each iteration, until an equilibrium point is reached [116]. In general, any implementation of this technique relies on two separate modules. These are the performance evaluation tool and the optimization tool [7]. Depending on the type of performance evaluation as well as the optimization technique used, different approaches can be distinguished [7]. With respect to performance evaluation, it is possible to use equations [117]. In this case, the optimization is known as equation-based. There also exists the possibility of evaluating the performance with the help of a simulation tool [118]. The optimization is then known as simulation-based. With respect to the optimization techniques, there exist two approximations: deterministic and statistical [7]. In EDA, examples of optimization-based sizing tools using simulation are DELIGHT.SPICE [118], MAELSTROM [119] and ANACONDA [120]. Examples of tools using equations are OPASYN [121] and STAIC [122]. When optimization-based synthesis is used for layout synthesis, two categories are usually considered. The first group is composed of heuristic-based approaches such as ILAC [123] and ANAGRAM [124]. The other group of approaches such as ROAD [125], is based on performance-driven optimization of the circuit layout.

CHAPTER 3

Behavioural Modelling of Chaotic Systems

Abstract: The goal of this book consists in the improvement of the EDA process for chaotic systems by means of the definition, development, and demonstration of an automatic synthesis methodology, which helps to reduce the design complexity for this class of systems and allows exploiting the possibilities of chaotic systems in future applications (e.g. chaos-based information systems). For this, it is necessary to properly define the level of abstraction used in the proposed synthesis approach to take full advantage of behavioural modelling as reviewed in the previous chapter. Therefore, the chaotic systems are modelled herein at the highest level of abstraction (ESL), by applying state variables approach and PWL approximations. In particular, the following chaotic systems are considered: Chua's circuit, Generalized Chua's circuit and a chaotic oscillator implemented using saturated functions. Further, time-efficient evaluation of chaotic behaviour, via numerical simulation, is presented by using multistep algorithms, from which chaotic dynamical behaviours and basic dynamical properties can be explored. The simulation is executed by automatic determination and control of step-size based on calculating the minimum eigenvalue of the state variables. Regarding to circuit implementation it is necessary to consider the real limitations of the electronic devices, therefore, two procedures (excursion levels scaling and frequency scaling) to modify the behaviour of chaotic systems are also introduced. Furthermore, by applying state variables and PWL approximations one can also simulate 2D and 3D scrolls attractors as shown in this chapter.

Keywords: Chaos, electronic design automation, dynamical systems, chaos generators, modelling and simulation.

STATE VARIABLES AND PWL APPROXIMATION

The behavioural model of a nonlinear dynamical system can be written in terms of state space representation as given in (2.1) [31,90], which consist of a set of differential equations describing the evolution of the variables. These are known as the state variables and their values at any particular time are supposed to contain sufficient information for the future evolution of the system. Since there is one equation for each state variable, the order of the system is thus the number of independent equations and hence also the number of independent initial conditions [102]. For autonomous chaotic systems, the state variables are defined by (1), where the function $\mathbf{f}$ is nonlinear. Therefore, the piecewise-linear (PWL) approximation is used herein to describe this nonlinear function, which consists of a set of linear relations valid in different regions [103,104]. Such functions are not analytic at all points, since they contain discontinuities of value or gradient, but they have the advantage that the dynamical equations become linear (and hence soluble) in any particular region, and the solutions for different regions can then be joined together at the boundaries [104]. Furthermore, (1) can be described by a linear state-space representation given in (2), where $\mathbf{A}$ and $\mathbf{B}$ are matrices of appropriate dimensions. The great advantage of linearity is that, even in the time dependent case, a formal solution can immediately be constructed, which is moreover applicable for all initial conditions and all input functions [102-104, 126].

$$\dot{\mathbf{x}} = \mathbf{f}(\mathbf{x}) \tag{1}$$

$$\dot{\mathbf{x}} = \mathbf{A}\mathbf{x} + \mathbf{B}\mathbf{u} \tag{2}$$

As a result, it is possible to give an essentially complete classification of behaviour in the phase plane [26]. In general, the equations describing a nonlinear dynamical system cannot be solved analytically, so that, in order to construct the trajectories accurately, it is necessary to use numerical methods [33-35]. The next sections describe the behavioural modelling of chaotic oscillators by state-variables and PWL approach.

Chua's Circuit

Chua's circuit is a chaotic system which can be easily built, simulated, and tractable mathematically [28-31]. It consists of five circuit elements: one linear resistor (R), one inductor, two capacitors, and one nonlinear resistor known as Chua's diode. Chua's circuit can be modelled by applying the state variables approach [28,126]. In this manner, the behavioural modelling from Fig. **1**(a) consists of three equations, i.e. one for each state variable, as shown by (3)-(5). In (3), i_{NR} describes the current through Chua's diode, which is a nonlinear resistor. The main idea behind is to transform the problem of solving a nonlinear dynamical system of differential equations into a sequence of linear and purely algebraic problem which can be solved straightforward by applying numerical simulation [33]. Furthermore, Chua's diode can be modelled by an I-V piecewise-linear characteristic, as shown by Fig. **1**(b). As one sees, for voltage signals less than the breakpoint (*BP1*) in absolute value the characteristic has a linear segment with negative slope *g1*. For absolute voltages larger than *BP1* the I-V PWL characteristic has two linear segments of negative slope *g2*. The negative PWL behaviour is valid in the nominal range (*-BP2, BP2*), in which the diode is normally operated [29]. For voltages outside this range the slope of the I-V PWL characteristic increases monotonically ultimately becoming positive *g3*. In this manner, i_{NR} has the general form given by (6), where *m* (*g1, g2, g3*) and *Ix* should be updated according to (7).

$$\frac{dV_{C1}}{dt} = -\frac{V_{C1}}{RC_1} + \frac{V_{C2}}{RC_1} - \frac{i_{NR}}{C_1} \tag{3}$$

$$\frac{dV_{C2}}{dt} = \frac{V_{C1}}{RC_2} - \frac{V_{C2}}{RC_2} - \frac{i_L}{C_2} \tag{4}$$

$$\frac{dI_L}{dt} = -\frac{V_{C2}}{L} \tag{5}$$

$$i_{NR} = mV_{C1} + Ix \tag{6}$$

$$i_{NR} = \begin{cases} -g2V_{C1} + (g1-g2)BP1 & V_{C1} < -BP1 \\ -g1V_{C1} & -BP1 \le V_{C1} \le BP1 \\ -g2V_{C1} + (g2-g1)BP1 & V_{C1} > BP1 \end{cases} \tag{7}$$

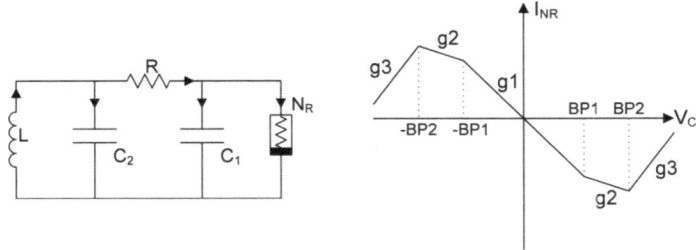

Figure 1: Chua's circuit and driving-point I-V piecewise-linear characteristic of Chua's diode.

Equations (3)-(7) lead to three systems of first-order differential equations, as shown by (8)-(10).

$$\begin{bmatrix} \dot{V}_{C1} \\ \dot{V}_{C2} \\ \dot{I}_L \end{bmatrix} = \begin{bmatrix} -\frac{1}{RC_1} + \frac{g2}{C_1} & \frac{1}{RC_1} & 0 \\ \frac{1}{RC_2} & -\frac{1}{RC_2} & \frac{1}{C_2} \\ 0 & -\frac{1}{L} & 0 \end{bmatrix} \begin{bmatrix} V_{C1} \\ V_{C2} \\ I_L \end{bmatrix} + \begin{bmatrix} \frac{(g2-g1)BP1}{C_1} \\ 0 \\ 0 \end{bmatrix} V_{C1} < -BP1 \tag{8}$$

$$\begin{bmatrix} \dot{V}_{C1} \\ \dot{V}_{C2} \\ \dot{I}_L \end{bmatrix} = \begin{bmatrix} -\frac{1}{RC_1} + \frac{g1}{C_1} & \frac{1}{RC_1} & 0 \\ \frac{1}{RC_2} & -\frac{1}{RC_2} & \frac{1}{C_2} \\ 0 & -\frac{1}{L} & 0 \end{bmatrix} \begin{bmatrix} V_{C1} \\ V_{C2} \\ I_L \end{bmatrix} - BP1 \le V_{C1} \le BP1 \tag{9}$$

$$
\begin{bmatrix} \overset{\bullet}{V_{C1}} \\ \overset{\bullet}{V_{C2}} \\ \overset{\bullet}{I_L} \end{bmatrix} = \begin{bmatrix} -\dfrac{1}{RC_1} + \dfrac{g2}{C_1} & \dfrac{1}{RC_1} & 0 \\ \dfrac{1}{RC_2} & -\dfrac{1}{RC_2} & \dfrac{1}{C_2} \\ 0 & -\dfrac{1}{L} & 0 \end{bmatrix} \begin{bmatrix} V_{C1} \\ V_{C2} \\ I_L \end{bmatrix} + \begin{bmatrix} \dfrac{(g1-g2)BP1}{C_1} \\ 0 \\ 0 \end{bmatrix} V_{C1} > BP1
\tag{10}
$$

Generalized Chua's Circuit

In [65] is shown the generalization of Chua's circuit whose behavioural equations are given by (11), where $h(x)$ describes a PWL characteristic given by (12), where q is a natural number adjusted to generate even or odd-scrolls attractors, m are the slopes of the PWL characteristic, and c are the breaking points [127]. The values of m and c for $q=1,2,3$, have been determined in [65].

$$
\dot{x} = \alpha\left[y - h(x)\right] \qquad \dot{y} = x - y + z \qquad \dot{z} = -\beta y
\tag{11}
$$

$$
h(x) = \left(m_{2q-1}\right)x + \frac{1}{2}\sum_{i=1}^{2q-1}\left(m_{i-1} - m_i\right) \times \left(\left|x + c_i\right| - \left|x - c_i\right|\right)
\tag{12}
$$

In Fig. **2** is shown the generalized Chua's circuit implemented with a voltage-controlled voltage source (VCVS) [65]. The state-variables equations are derived in (13). By setting C1=C2=R=RG=L=1, and by choosing $V_{C1} = x$, $V_{C2} = y$, $I_L = z$, (13) can be described by (14), where $f(x) = -h(x) + (1+\delta)x$ and it depends on the number of slopes and breaking points of the nonlinear function $h(x)$ in (12). By selecting $q = 2$, $m = [0.9/7, -3/7, 3.5/7, -2.4/7]$, and $c = [1; 2.15; 4]$, the PWL graph of (12) is shown in Fig. **3**(a). If $q = 3$, $m = [0.9/7, -3/7, 3.5/7, -2.7/7, 4/7, -2.4/7]$, and $c = [1; 2.15; 3.6; 6.2; 9]$, the PWL graph is shown in Fig. **3**(b).

$$
\frac{dV_{C1}}{dt} = -\frac{V_{C1}}{RR_GC_1}(R + R_G) + \frac{V_{C2}}{RC_1} + \frac{f(V_{C1})}{C_1} \quad \frac{dV_{C2}}{dt} = \frac{V_{C1}}{RC_2} - \frac{V_{C2}}{RC_2} + \frac{I_L}{C_2} \quad \frac{dI_L}{dt} = -\frac{V_{C2}}{L}
\tag{13}
$$

$$
\begin{bmatrix} \overset{\bullet}{x} \\ \overset{\bullet}{y} \\ \overset{\bullet}{z} \end{bmatrix} = \begin{bmatrix} -\alpha(1+\delta) & \alpha & 0 \\ 1 & -1 & 1 \\ 0 & -\beta & 0 \end{bmatrix} \begin{bmatrix} x \\ y \\ z \end{bmatrix} + \begin{bmatrix} \alpha f(x) \\ 0 \\ 0 \end{bmatrix}
\tag{14}
$$

Multi-scrolls Chaotic Oscillator

The previous chaotic oscillators could generate multi-scrolls with one-dimensional (1D) orientation, however in this subsection is shown a chaotic oscillator, which can generate scrolls with 2D and 3D orientations. Indeed, a chaotic oscillator implemented with saturated circuits is well suited to implement PWL approximations [79]. For instance, the behaviour of the opamp can be modelled by using saturated functions [20,96]. In Fig. **4** are shown the saturated functions with five-segments and seven segments, respectively. Two different segments compose these functions, the segments without slope are called saturated plateaus and the segments with slope are called saturated slopes. Thereby, the number of scrolls that these SFs can generate depends on the number of saturated plateaus [20]. In (15) is described a PWL approximation called series of a saturated function [79], which is shown in Fig. **4**, where $k > 0$ is the slope of the saturated function, $h > 2$ is the saturated delay, p and q are positive integers. One can recast (15) in an explicit form as shown in (16).

$$
f(x; k, h, p, q) = \sum_{i=-p}^{q} f_i(x; k, h)
\tag{15}
$$

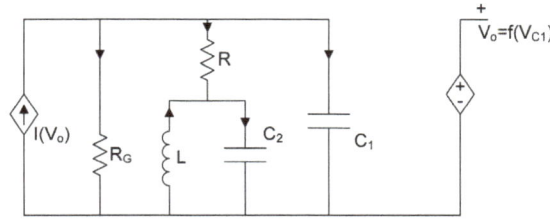

Figure 2: Generalized Chua's circuit using a VCVS.

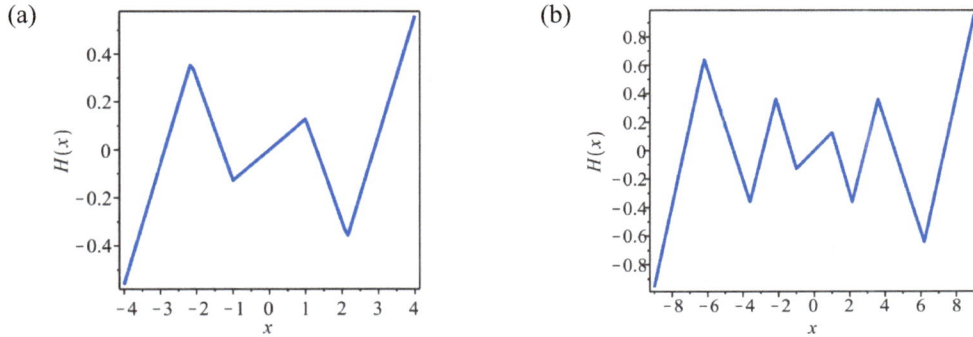

Figure 3: PWL approximation of *h(x)*: (a) for *q*=2, and (b) for *q*=3.

$$f(x;k,h,p,q) = \begin{cases} (2q+1)k & x > qh+1 \\ k(x-ih)+2ik & |x-ih| \le 1, -p \le i \le q \\ (2i+1)k & ih+1 < x < (i+1)h-1, -p \le i \le q-1 \\ -(2p+1)k & x < -ph-1 \end{cases} \tag{16}$$

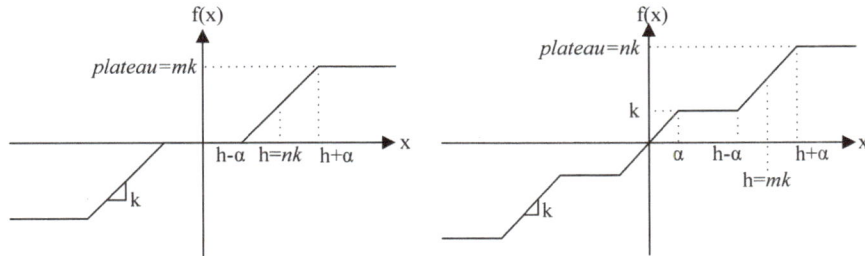

Figure 4: PWL description of a series of saturated functions to generate odd and even-scrolls.

1D-multi-scroll Chaotic Attractors

To generate multi-scrolls attractors with one-dimensional (1D) orientation, a controller is added to a system of state-variables equations [79], as shown in (17), where $f(x;k,h,p,q)$ is defined by (16) and x, y, z are state variables and a, b, c, d are positive constants.

$$\dot{x} = y \qquad \dot{y} = z \qquad \dot{z} = -ax - by - cz + df(x;k,h,p,q) \tag{17}$$

In (17) there are $2(p+q)+3$ equilibrium points on the x-axis, called saddle points of index 1 and index 2. Since the scrolls are generated only around the saddle points of index 2 [24], (17) has the potential to generate $(p+q+2)$-scrolls for some suitable parameters a, b, c, d, k, h. However, the $(p+q+1)$ saddle points of index 1 are responsible to connect the $(p+q+2)$-scrolls to generate the attractor. Additionally, the saddle points of index 2 correspond to a unique saturated plateau and to a unique attractor, while the saddle points of index 1 correspond to a unique saturated slope and connection among the neighbors-scrolls [128].

Besides, the saturated plateau in saturated function series in (16) is: $plateau = \pm nk$ for even-scrolls and $plateau = \pm mk$ for odd-scrolls. The saturated delays for the slopes centers are defined by $h_i = \pm mk$ for even-scrolls and $h_i = \pm nk$ for odd-scrolls as shown in Fig. **4**. The multiplier factors for the above-mentioned expressions are defined by $n = 1,3...(p+q+1)$ for even-scrolls and $n = 1,3...(p+q-1)$ for odd-scrolls; and $m = 2,4...(p+q)$ for both types of scrolls. It can be concluded that the position of the multi-scrolls in the chaotic attractor depends completely on the value of k as shown in the next section.

2D-multi-scroll Chaotic Attractors

The chaotic system in (17) is modified to generate chaotic behaviour on a 2D-mesh [20,127]. The 2D chaotic system is modelled by applying state variables approach as shown in (18), where x, y, z are state variables, and a,b,c, d_1,d_2 are positive real constants. Two saturated function series $f(x)$ and $f(y)$ in (18) are needed to generate 2D-multi-scrolls attractors and are also defined by (16), where p_1, p_2, q_1 and q_2 are positive integers [79]. Therefore, the chaotic system has the potential to create a 2D $(p_1 + q_1 + 2) \times (p_2 + q_2 + 2)$-even-scrolls mesh and a 2D $(p_1 + q_1 + 1) \times (p_2 + q_2 + 1)$-odd-scrolls mesh for suitable parameters a,b,c,d_1,d_2, k_1, k_2, h_1 h_2. Besides, the saturated plateau in saturated function series in (16) is: $plateau = \pm nk$ for 2D-even-scrolls and $plateau = \pm mk$ for 2D-odd-scrolls. The saturated delays for the slopes centers are defined by $h_i = \pm mk$ for 2D-even-scrolls and $h_i = \pm nk$ for 2D-odd-scrolls as shown in Fig. **1**. The multiplier factors for the above-mentioned expressions are defined by $n = 1,3...(p_2 + q_2 + 1)$ for 2D-even-scrolls and $n = 1,3...(p_2 + q_2 - 1)$ for 2D-odd-scrolls; and $m = 2,4...(p_1 + q_1)$ for both types of scrolls.

$$\dot{x} = y - \frac{d_2}{b} f(y; k_2, h_2, p_2, q_2) \qquad \dot{y} = z \qquad \dot{z} = -ax - by - cz + d_1 f(x; k_1, h_1, p_1, q_1) + d_2 f(y; k_2, h_2, p_2, q_2) \qquad (18)$$

Additionally, the centers of scrolls and connections among neighbors-scrolls in a 2D-scrolls mesh depend completely on the value of k. This book proposes that both centers and connections are evaluated by matrix representations shown from (19) to (23). The matrixes are filled in a (x,y) form, where x and y are the values on the x-axis and y-axis, respectively. All scrolls have a radius of k. In (20) and (23), the operation (*) means an interchange in the axis as shown here $(x,y) \rightarrow (*) = (y,x)$. Also, one needs to evaluate all quadrants from (19) to (23), this is $(+x,+y),(-x,+y),(-x,-y),(+x,-y)$, for simplicity they are only given in the first quadrant $(+x,+y)$. Consequently, centers of the 2D even-scrolls are defined by $\mathbf{C}$ matrix in (19); the connections are defined by $\mathbf{U}_x$, $\mathbf{U}_y$, $\mathbf{U}$, and $\mathbf{U}'$ matrices in (20) and (23). For 2D odd-scrolls, centers of scrolls are defined by $\mathbf{C}'$ matrix in (21) and the connections are defined by $\mathbf{U}'_x$, $\mathbf{U}'_y$, $\mathbf{U}$, and $\mathbf{U}'$ matrices in (22) and (23). Similarly, one can design 2D-multi-scrolls attractors in (x,z) or (y,z) directions.

$$\mathbf{C} = \begin{bmatrix} (k,k) & \cdots & (nk,k) \\ \vdots & \ddots & \vdots \\ (k,nk) & \cdots & (nk,nk) \end{bmatrix} \qquad (19)$$

$$\mathbf{U}_x = \begin{bmatrix} (k,2k) & \cdots & (nk,2k) \\ \vdots & \ddots & \vdots \\ (k,mk) & \cdots & (nk,mk) \end{bmatrix} \quad \mathbf{U}_y = \mathbf{U}_x{}^{\mathbf{T}}(*) = \begin{bmatrix} (2k,k) & \cdots & (mk,k) \\ \vdots & \ddots & \vdots \\ (2k,nk) & \cdots & (mk,nk) \end{bmatrix} \qquad (20)$$

$$\mathbf{C}' = \begin{bmatrix} (0,0) & \cdots & (mk,0) \\ \vdots & \ddots & \vdots \\ (0,mk) & \cdots & (mk,mk) \end{bmatrix} \qquad (21)$$

$$\mathbf{U}'_x = \mathbf{U}_y{}^{\mathbf{T}} = \begin{bmatrix} (2k,k) & \cdots & (2k,nk) \\ \vdots & \ddots & \vdots \\ (mk,k) & \cdots & (mk,nk) \end{bmatrix} \quad \mathbf{U}'_y = \mathbf{U}_x{}^{\mathbf{T}} = \begin{bmatrix} (k,2k) & \cdots & (k,mk) \\ \vdots & \ddots & \vdots \\ (nk,2k) & \cdots & (nk,mk) \end{bmatrix} \qquad (22)$$

$$\mathbf{U} = \begin{bmatrix} (k,0) & \cdots & (nk,0) \end{bmatrix} \quad \mathbf{U}' = \mathbf{U}(*) = \begin{bmatrix} (0,k) & \cdots & (0,nk) \end{bmatrix} \qquad (23)$$

3D-multi-scroll Chaotic Attractors

The chaotic system in (17) is modified in a similar form to (18) to generate chaotic behaviour on a 3D-mesh. The 3D chaotic system is again modelled by applying state variables approach [89] as shown in (24), where x, y, z are state variables, and a, b, c, d_1, d_2, d_3 are positive real constants. Three saturated function series $f(x), f(y)$ and $f(z)$ in (24) are needed to generate 3D-multi-scrolls attractors and are also defined by (16), where p_1, p_2, p_3, q_1, q_2, q_3 are positive integers [79]. Therefore, the chaotic system has the potential to create a 3D $(p_1 + q_1 + 2)$ x $(p_2 + q_2 + 2)$ x $(p_3 + q_3 + 2)$-even-scrolls mesh and a 2D $(p_1 + q_1 + 1)$ x $(p_2 + q_2 + 1)$ x $(p_3 + q_3 + 1)$-odd-scrolls mesh for suitable parameters $a, b, c, d_1, d_2, d_3, k_1, k_2, k_3, h_1, h_2, h_3$. The value of saturated plateaus and saturated delays in saturated function series in (16) obeys to the same rules as described in the previous case.

$$\dot{x} = y - \frac{d_2}{b} f(y; k_2, h_2, p_2, q_2) \qquad \dot{y} = z - \frac{d_3}{c} f(z; k_3, h_3, p_3, q_3)$$

$$\dot{z} = -ax - by - cz + d_1 f(x; k_1, h_1, p_1, q_1) + d_2 f(y; k_2, h_2, p_2, q_2) + d_3 f(z; k_3, h_3, p_3, q_3)$$

$$(24)$$

Besides, the centers of the scrolls and their connections can also be evaluated by the matrix representations from (19) to (23). However, it should be pointed out that these evaluations must be done for each phase-plane in (x, y), (y, z) and (z, x) directions. Therefore, the centers of the 3D even-scrolls are defined by $\mathbf{C}$ matrix in (19); the connections are defined by $\mathbf{U}_x$, $\mathbf{U}_y$, $\mathbf{U}$, and $\mathbf{U}'$ matrices in (20) and (23). For 3D odd-scrolls, centers of scrolls are defined by $\mathbf{C}'$ matrix in (21) and the connections are defined by $\mathbf{U}'_x$, $\mathbf{U}'_y$, $\mathbf{U}$, and $\mathbf{U}'$ matrices in (22) and (23). One should note that subindex x and y in (20), (22) can be changed to (y, z) or (z, x).

In the previous sections, it has been shown the behavioural modelling of chaotic systems using state variables and piecewise linear approach. The next section introduces a method to automatically control and determine the step-size for a multi-step algorithm leading to numerical simulation results for the three kinds of chaotic oscillators.

AUTOMATIC NUMERICAL INTEGRATION ALGORITHM

The analysis of chaotic dynamical systems is one of the most challenging tasks in computational science because these systems are essentially nonlinear [33]. Besides, their behaviour is much more complicated than that of linear systems. In fact, even the simplest chaotic system exhibit a bulk of different behaviours that only can be fully analyzed with the help of powerful software resources [32]. In this manner, this section is focused to introduce an automatic algorithm to simulate the behaviour of chaotic systems using the high-level modelling approach introduced in the previous chapter. The main idea behind is to transform the problem of solving a nonlinear dynamical system described by differential equations into a sequence of linear and purely algebraic problem which can be solved straightforward by applying numerical simulation to compute a sequence of chaotic phenomena by solving the state-variables equations. Consequently, one can explore the space design by observing the corresponding effects on and the relationships between the chaotic system parameters, so that the synthesis of the PWL-function-based chaotic systems can be realized directly using electronic devices. There are two different approaches for numerical algorithms, Taylor series based-algorithms and polynomial based-algorithms [102]. In the former, the explicit fourth-order Runge-Kutta (RK) method is the most widely used for large step-size. In the latter, the explicit Adams-Bashfort algorithms and implicit Adams-Moulton algorithms are the most frequently used for establishing adaptive order and step-size schemes. An important thing is that the local truncation error for implicit algorithms is very small than for explicit algorithms [34]. In this manner, the application of implicit multistep algorithms based on polynomial approximation is presented to simulate the behaviour of the chaotic systems [126,129].

Henceforth, the Forward Euler (FE) approximation [35] given by (25) is used to start the Adams-Moulton (AM) algorithms, which are implicit multistep ones. For instance, the third-order AM algorithm [102] is given by (26), where the term $f(x_{n+1}, t_{n+1})$ can be predicted using (25). The evaluation of behavioural representation of the chaotic systems given by the state variables $\dot{x} = \mathbf{A}x + \mathbf{B}u$, using this implicit algorithm derives (27), where $\mathbf{A}$ and $\mathbf{B}$ are the state-matrix and PWL characteristics of the chaotic systems shown in section 3.1. h, x_n and x_{n-1} are the step-size and state vectors evaluated at one and two past steps, and $\hat{x}_{n+1}$ is predicted using FE. The local error truncation is given by (28) [34,102].

$$x_{n+1} = x_n + hf(t_n, x_n)$$ (25)

$$x_{n+1} = x_n + h\left\{\frac{5}{12}f(x_{n+1}, t_{n+1}) + \frac{8}{12}f(x_n, t_n) - \frac{1}{12}f(x_{n-1}, t_{n-1})\right\}$$ (26)

$$x_{n+1} = x_n + h\left\{\frac{5}{12}(A\hat{x}_{n+1} + B) + \frac{8}{12}(Ax_n + B) - \frac{1}{12}(Ax_{n-1} + B)\right\}$$ (27)

$$\varepsilon_T = \left[C_k \hat{x}^{(k+1)}(\hat{\tau})\right]h^{k+1} = \left[-\frac{1}{24}\hat{x}^{(4)}(\hat{\tau})\right]h^4$$ (28)

The next two sections introduce the proposed approaches to select and control the step-size. For sake on simplicity, they are firstly shown for simulating Chua's circuit and afterwards, these can be extended to other kinds of chaotic systems, e.g. multi-scrolls chaotic oscillators.

AUTOMATIC CONTROL OF STEP-SIZE

If a numerical method is selected, e.g. the third-order AM, and its order does not change during the integration process, then the step-size can be optimized by selecting the highest possible value for which the local error is limited below a maximum error given by $h \cdot e_{max}$ [34], to guarantee the stability of the numerical method. The local error truncation can be described by (29), where k denotes the order of the algorithm. By assuming that $|\varepsilon_T| = h \cdot e_{max}$, then e_{max} can be described by (30), which shows the relationship between the maximum values of e_{max} and h. On the other hand, the value of h before the calculation of the next value for $x(t)$, can be updated or modified to be $\overset{\Delta}{h} = \alpha h$, where α is a constant of proportionality [33]. Now, the local error truncation in $t = t_n$ for the AM algorithms is given by (31), where the gradient is evaluated by (32). Furthermore, by solving for α in (31), one gets (33). For instance, for the third-order AM algorithm with $C_3 = 1/24$ and $k=3$ [102], (33) results in (34).

$$\varepsilon_T = \left[C_k x^{(k+1)}\right]h^{k+1} = O(h^{k+1})$$ (29)

$$e_{max} = \left[C_k x^{(k+1)}\right]h^k$$ (30)

$$\varepsilon_T \cong C_k \alpha^{k+1}(k!)\left[\nabla^1(z_n)_{k+1}\right]$$ (31)

$$\nabla^1(z_n)_{k+1} = \frac{h^{k+1}x^{(k+1)}}{k!}$$ (32)

$$\alpha \cong \left\{\frac{\varepsilon_T}{C_k(k!)[\nabla^1(z_n)_{k+1}]}\right\}^{\frac{1}{k+1}} = \left\{\frac{\varepsilon_T}{C_k h^{k+1}x^{(k+1)}}\right\}^{\frac{1}{k+1}}$$ (33)

$$\alpha \cong \left\{\frac{24\varepsilon_T}{h^4 x^{(4)}}\right\}^{\frac{1}{4}}$$ (34)

In this case, to adjust h, one can use α_1 and α_2, which are functions of e_{max} and are given by (35) and (36). By setting the intervals for $\alpha_1 = \{0, \varepsilon_i\}$ and $\alpha_2 = \{\varepsilon_s, \varepsilon_{max}\}$ the algorithm can be initialized [129]. At each iteration the algorithm evaluates that $\varepsilon_T \leq \varepsilon_{max}$, if the condition is not satisfied, then the value of the step size is not accepted. It is worthy to mention that multistep algorithms do not allow high values of h, because it causes convergence problems when α is very high [35].

$$\alpha_1 \cong \frac{1}{57}\left\{\frac{24\varepsilon_i}{h^4 x^{(4)}}\right\}^{\frac{1}{4}}$$ (35)

$$\alpha_2 \cong \frac{1}{18}\left\{\frac{24\varepsilon_s}{h^4 x^{(4)}}\right\}^{\frac{1}{4}} \tag{36}$$

Simulation of Chua's Circuit with Automatic Control of Step-Size

In Fig. **5** is shown the chaotic behaviour of Chua's circuit using the third-order AM algorithm described in the previously section by automatic control of step-size and without control of it. By setting: C1=450pF, C2=1.5nF, L=1mH, g1=1/1358, g2=1/2464, g3=1/1600, BP1=0.114V, BP2=0.4V, R=1625 in (8)-(10) and with h=0.1e-6, x_0=0.01, ε_i=1.1e-31, ε_s=1.2e-31 and e_{max}=2.5e-31 in (35)-(36) for initializing the algorithm, which are step-size, initial condition, inferior-error, superior-error, and maximum error, respectively, so that $\alpha_1 \cong 0.7$ and $\alpha_2 \cong 2.3$. The error generated by Fig. **5** is shown in Fig. **6**, where it can be appreciated that $\varepsilon_T \le \varepsilon_{max}$. In Fig. **7** are shown the values of h at each iteration. Finally, in Table **1** is shown that the AM algorithm with automatic control of step-size, using a PC Pentium 4 at 3GHz, leads us to a speed-up in time simulation.

Table 1: Comparison of the simulation of Chua's circuit using automatic control of step-size.

Algorithm	Iterations	Time(s)	Bytes	Final Time(s)
Third-order AM without automatic control of h	10,000	43.094	509196612	1e-3
Third-order AM with automatic control of h	5,152	7.922	171011588	1e-3

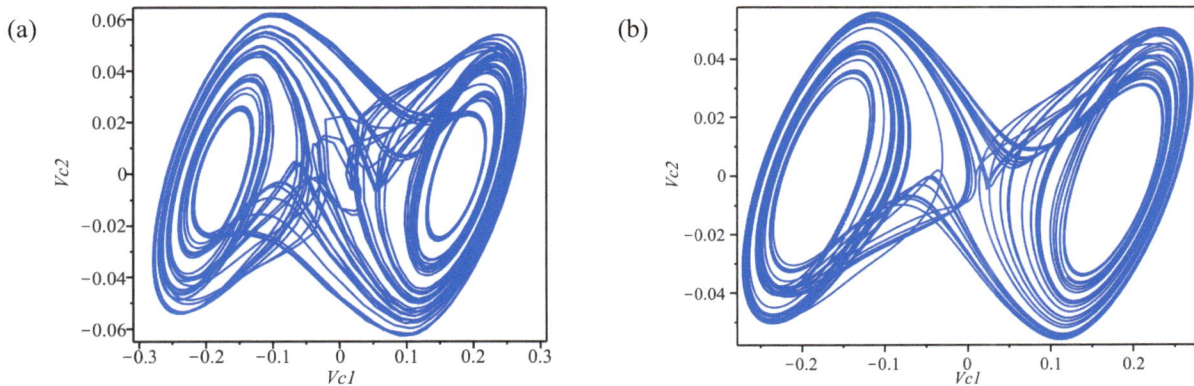

Figure 5: Simulation of Chua's circuit using the third-order AM algorithm: Double-scroll behaviour (a) with automatic control of h (5152 iterations), and (b) without control of h (10,000 iterations).

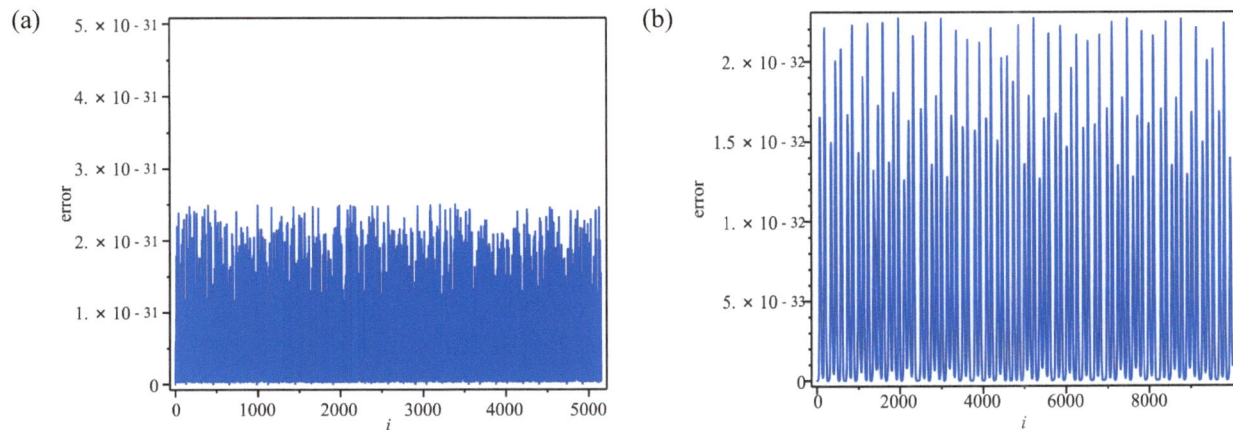

Figure 6: Error generated from: (a) Fig. **5**(a), and (b) Fig. **5**(b).

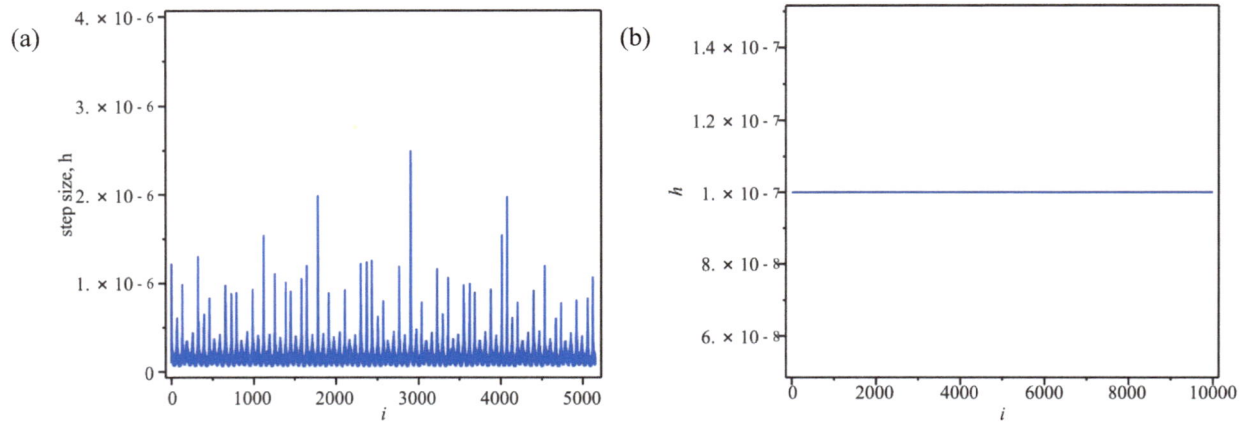

Figure 7: Values of the step-size from: (a) Fig. **5**(a), and (b) Fig. **5**(b).

AUTOMATIC DETERMINATION OF INITIAL STEP-SIZE

In the previous section it was shown that if the order of the numerical method (for instance, the third-order AM algorithm) does not change during the integration process, then h can be optimized by selecting the highest possible value for which the local error is limited below a maximum-error given by $h \cdot e_{max}$, to guarantee the stability of the numerical method. However, to initialize the algorithm in (27) with automatic control of step size, it is necessary to give the initial values for initial step-size h, inferior-error, superior-error, and maximum error in (35) and (36). In this manner, a procedure to speed-up time simulation of chaotic systems in a full automatic fashion is introduced in this section.

The multistep method is applied with automatic control of step-size h as shown in the previous section. However; the initial value of h along with the necessary values to initialize time simulation are herein determined by the computation of the eigenvalues of the state matrices of the chaotic systems which change according to the slopes of the PWL functions that describe the nonlinear behaviour of the chaotic systems. The determination of the eigenvalues of a system can be computed by evaluating (37), where **A** is the state matrix, **I** is the identity matrix and Δ express the determinant of $|\mathbf{A}-\lambda\mathbf{I}|$. Eq. (37) is known as the characteristic equation of **A** [26], and its evaluation generates a characteristic polynomial, from which are calculated the eigenvalues.

$$\Delta\left(\mathbf{A}-\lambda\mathbf{I}\right)=0 \tag{37}$$

The initial h for the third-order AM algorithm in (27), was given by the user and if the values of the circuit components, parameters, slopes and/or breakpoints of the chaotic systems change, the user needs to estimate a new initial step-size. Furthermore, the necessary values to initialize the simulation with automatic control of h, such as: inferior-error, superior-error and maximum-error, were also given by the user. In this manner, the initial h is computed from (38), where λ_{min} is the minimum absolute value of all eigenvalues expressed by (39). Ψ in (38) is estimated by applying the sample theorem, and therefore, its minimum value is 2.

$$h=\frac{\left(1/\lambda_{min}\right)}{\Psi} \tag{38}$$

$$\lambda_{min}=\min\left\{\left|\lambda_1\right|,\left|\lambda_2\right|,...,\left|\lambda_n\right|\right\} \tag{39}$$

The values of ε_i, ε_s and e_{max} are given by (40), (41) and (42), respectively.

$$\varepsilon_i=4.1(0.1^{K+1})(h^{K+1}) \tag{40}$$

$$\varepsilon_s=4.5(0.1^{K+1})(h^{K+1}) \tag{41}$$

$$e_{\max} = 94.6(0.1^{K+1})(h^{K+1}) \tag{42}$$

As a result, the number of operations is reduced with respect to the previous section, leading to a more gain to speed-up time simulation [130]. In addition, the proposed procedure enables the user to change the values of components, parameters, slopes and/or breakpoints of the state variables systems and PWL functions to explore the chaotic behaviour and dynamical properties for chaotic systems.

Simulation of Chua's Circuit with Automatic Determination of Step-Size

The evaluation of (37) for Chua's circuit in (8)-(10) results in the characteristic equations given by (43) and (44). Again, by setting: C1=450pF, C2=1.5nF, L=1mH, g1=1/1358, g2=1/2464, g3=1/1600 and R=1625, the evaluation of (43) and (44) generates the eigenvalues given by (45) and (46), respectively.

$$\lambda_A^3 + \frac{(LC_1 - g_1RLC_2 + LC_2)}{RLC_1C_2}\lambda_A^2 + \frac{(RC_1 - g_1L)}{RLC_1C_2}\lambda_A + \frac{(1 - g_1R)}{RLC_1C_2} = 0 \tag{43}$$

$$\lambda_B^3 + \frac{(LC_1 - g_2RLC_2 + LC_2)}{RLC_1C_2}\lambda_B^2 + \frac{(RC_1 - g_2L)}{RLC_1C_2}\lambda_B + \frac{(1 - g_2R)}{RLC_1C_2} = 0 \tag{44}$$

$$\lambda_{1A} = 522962.88 \quad \lambda_{2A,3A} = -332173.63 \pm 482094.83i \tag{45}$$

$$\lambda_{1B} = -920141.01 \quad \lambda_{2B,3B} = 22119.56 \pm 580416.45i \tag{46}$$

The evaluation of (39) results in $\lambda_{\min} = 522962.88$, and therefore (38) results in $h{\approx}0.12747877e{-}6$ with $\Psi = 19$, for instance. Furthermore, by solving (40), (41) and (42), one gets $\varepsilon_i \approx 1.1e{-}31$, $\varepsilon_s \approx 1.2e{-}31$ and $e_{\max} \approx 2.5e{-}31$. In Fig. **8** is shown the simulation of Chua's circuit with these values for comparing it with the simulations in Fig. **5**(a) and Fig. **7**(a).

Finally, in Table **2** is shown that the AM algorithm with automatic determination and control of step-size, using a PC Pentium 4 at 3GHz, leads us to speed-up in time simulation.

Table 2: Comparison of the simulation of Chua's circuit using automatic determination and control of step-size.

Algorithm	iterations	Time(s)	Mbytes	Final time(s)
Third-order AM without automatic control of h	10,000	43.094	509	1e-3
Third-order AM with automatic control of h	5,152	7.922	171	1e-3
Third-order AM with automatic determination and control of h (eigenvalues)	3049	2.734	75	1e-3

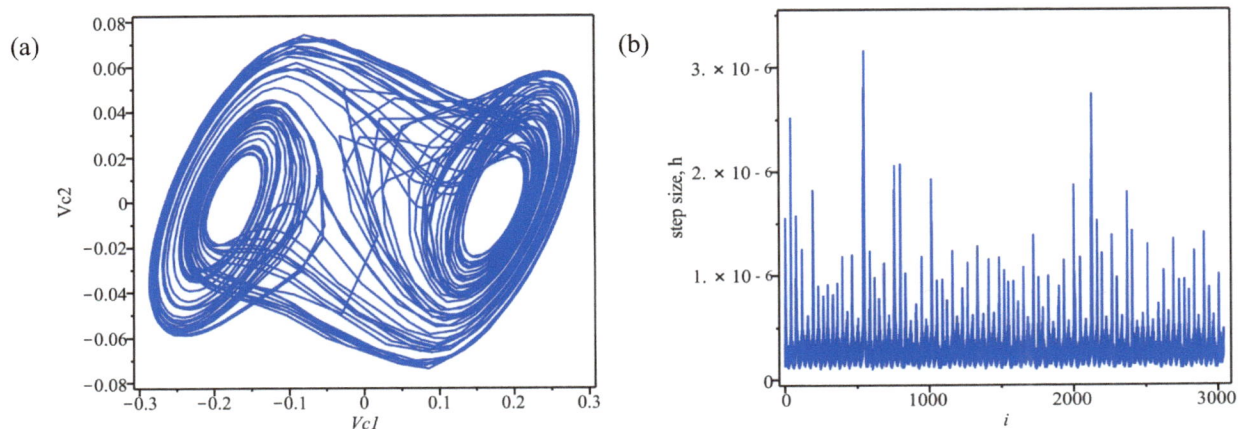

Figure 8: Simulation of Chua's circuit: (a) with automatic determination and control of h (3049 iterations), and (b) values of h.

NUMERICAL SIMULATION RESULTS

This section shows the numerical simulation results to generate 1D, 2D and 3D-multi-scrolls attractors. Basically, the numerical methods introduced in the previous sections are used herein to compute the eigenvalues and to control step-size automatically of three chaotic oscillators: Chua's circuit, generalized Chua's circuit, and multi-scrolls chaotic oscillator which have been modelled by using state variables and PWL approach. In this manner, this is the first step for the development of a synthesis approach to design chaotic oscillators, by beginning with high-level simulations, and ending with the synthesis of individual blocks using practical electronic devices.

Chua's Circuit

The numerical results for Chua's circuit were shown in the previous sections. In particular, Fig. **8**(a) shows the double-scroll behaviour.

Generalized Chua's Circuit

Chua's circuit can generate multi-scrolls chaotic attractors according to the number of segments in the PWL function. To generate 3 and 5-scrolls attractors from Fig. **2**, the simulation of (14) is performed by using the PWL approximations shown in Fig. **3**, respectively. The results are shown in Fig. **9** by selecting $q = 2$, $m = [0.9/7, -3/7, 3.5/7, -2.4/7]$, $c = [1; 2.15; 4]$, with $\alpha = 9$, $\beta = 14.28$ $\delta = 1$ (C1=1/9F, C2=1F, L=70mH, R=1Ω and RG=1Ω) in (12) and (14) and by selecting $q = 3$, $m = [0.9/7, -3/7, 3.5/7, -2.7/7, 4/7, -2.4/7]$, $c = [1; 2.15; 3.6; 6.2; 9]$, with $\alpha = 9$, $\beta = 14.28$ $\delta = 1$ in (12) and (14), respectively.

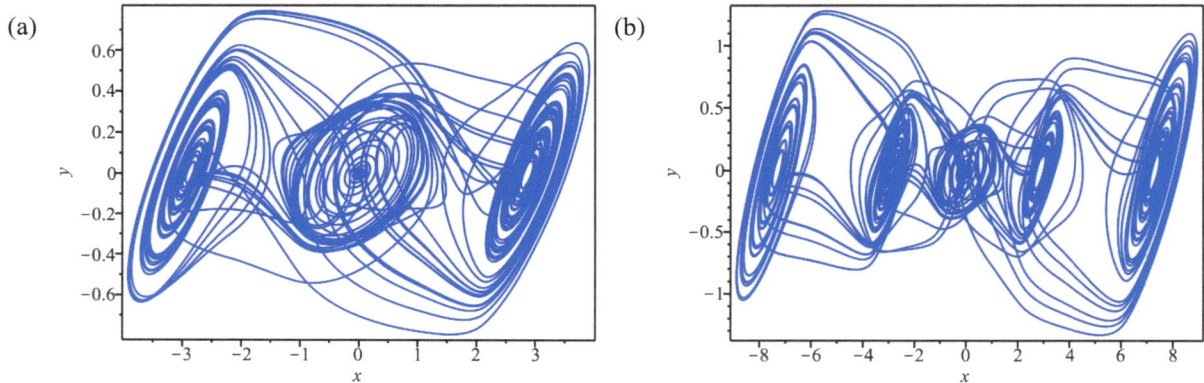

Figure 9: Generation of (a) 3-scrolls, and (b) 5-scrolls attractors with the generalized Chua's circuit.

Multi-scrolls Chaotic Oscillator

1D-multi-scroll Chaotic Attractors

By setting $a = b = c = d = 0.7$, $k = 10$, $h = 10$ and $h = 20$ to evaluate (17), 3-scroll and 6-scrolls attractors are generated, as shown in Fig. **10**, with $p = q = 1$ and $p = q = 2$, respectively.

2D-multi-scroll Chaotic Attractors

A 2D-3-scrolls chaotic attractor is generated by setting a=b=c=d1=d2=0.7, k1=k2=50, h1=h2=50, p1=q1=p2=q2=1 to evaluate (18) as shown in Fig. **11** and a 2D-4-scrolls attractor is generated with a=b=c=d1=d2=0.7, k1=k2=50, h1=h2=100, p1=q1=p2=q2=1, as shown in Fig. **12**.

3D-multi-scroll Chaotic Attractors

A 3D-3-scrolls chaotic attractor is generated by setting a=d1=0.7, b=c=d2=d3=0.8, k1=100, h1=100, k2=k3=50, h2=h3=50, p1=q1=p2=q2=p3=q3=1 to evaluate (24) as shown in Fig. **13** and a 3D-4-scrolls attractor is generated with a=d1=0.7, b=c=d2=d3=0.8, k1=100, h1=200, k2=k3=50, h2=h3=100, p1=q1=p2=q2=p3=q3=1, as shown in Fig. **14**.

(a)

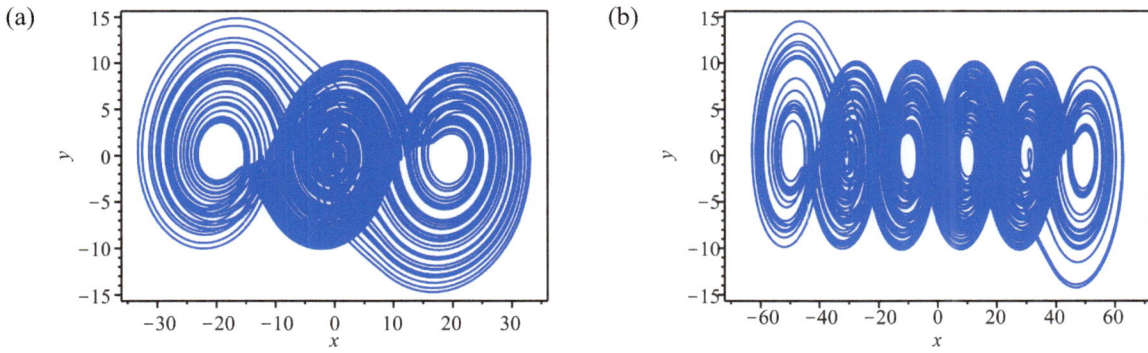

(b)

Figure 10: Generation of (a) 1D-3-scrolls, and (b) 1D-6-scrolls attractors.

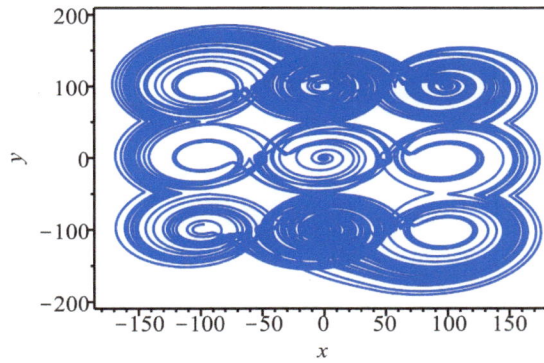

Figure 11: Generation of 2D-3-scrolls attractor.

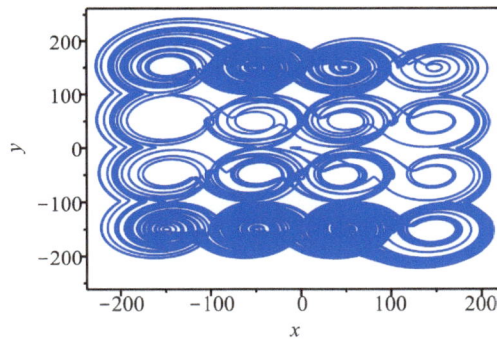

Figure 12: Generation of 2D-4-scrolls attractor.

(a)

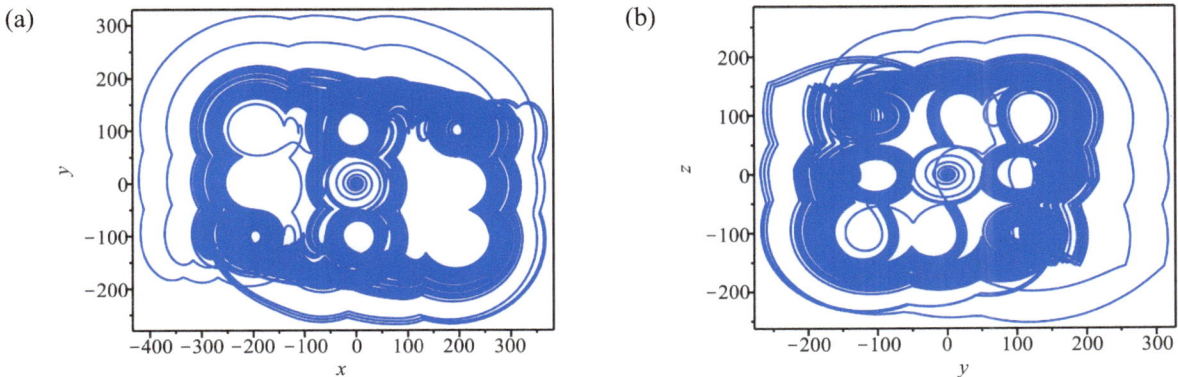

(b)

Figure 13: Generation of 3D-3-scrolls attractor: (a) x-y plane and (b) y-z plane.

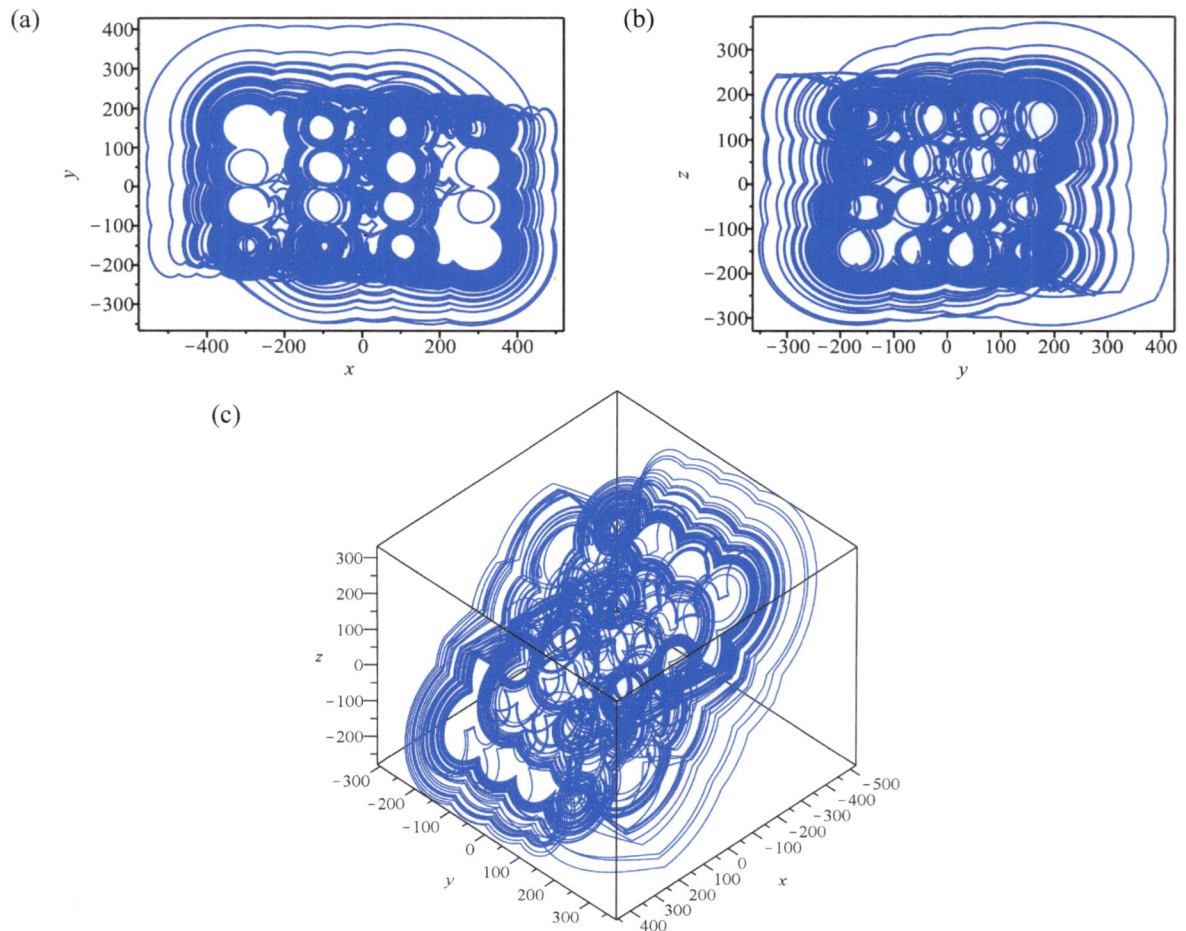

Figure 14: Generation of 3D-4-scrolls attractor: (a) x-y plane, (b) y-z plane and (c) its three-dimensional representation.

EXCURSION LEVEL SCALING OF CHAOTIC SIGNALS

Hardware implementation of reliable nonlinear circuits for generating various complex chaotic signals by using some simple electronic devices is a topic of both theoretical and practical interests [20]. However, as is well known, it is much more difficult to physically realize a nonlinear function that has an appropriate characteristic with many segments. Moreover, the realization of a nonlinear function with multi-segments is a basis for hardware implementation of chaotic attractors with multi-directional orientation and with a large number of scrolls [20]. One of the main obstacles is due to the limitation of the dynamic ranges of the available physical circuit components originating that the excursion of the signals is to be within these dynamic ranges [27]. Considering the aforementioned difficulties, a systematic procedure for circuit design is proposed in this section for scaling the excursion levels (ELs) of the PWL functions and consequently, the ELs of the chaotic signals.

The chaotic systems have been simulated by using behavioural modelling, from these numerical simulations one can see that the ELs of the chaotic attractors have wide values, especially the multi-scrolls chaotic attractor that real electronic devices cannot handle [128]. The basic idea is to add a parameter in the PWL functions. Thereby, one can control the ELs by selecting an appropriate scaling factor.

Chua's Circuit

To scale the EL of the Chua's circuit is necessary to add the parameter γ in the PWL function in (7) which describes the I-V characteristic of Chua's diode. Therefore, one can recast (7) by (47). The PWL function of Fig. **8** and PWL function with scaling factor of 100 are shown in Fig. **15**, respectively. The chaotic behaviour of the Chua's circuit is not affected by this scaling as shown in Fig. **16**.

$$i_{NR} = \begin{cases} -g2V_{C1} + (g1-g2)\left(\dfrac{BP1}{\gamma}\right) & V_{C1} < (-BP1/\gamma) \\ -g1V_{C1} & (-BP1/\gamma) \le V_{C1} \le (BP1/\gamma) \\ -g2V_{C1} + (g2-g1)\left(\dfrac{BP1}{\gamma}\right) & V_{C1} > (BP1/\gamma) \end{cases} \qquad (47)$$

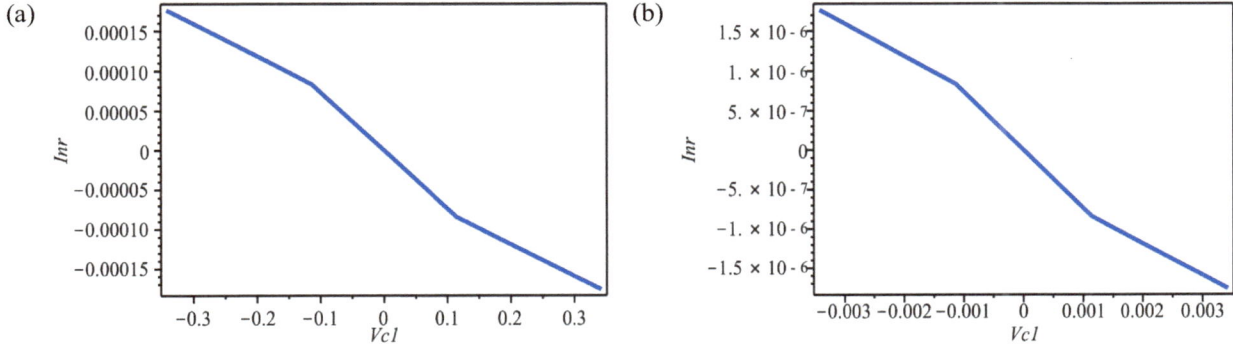

Figure 15: PWL function of Chua's circuit: (a) without scaling and (b) with scaling factor of 100.

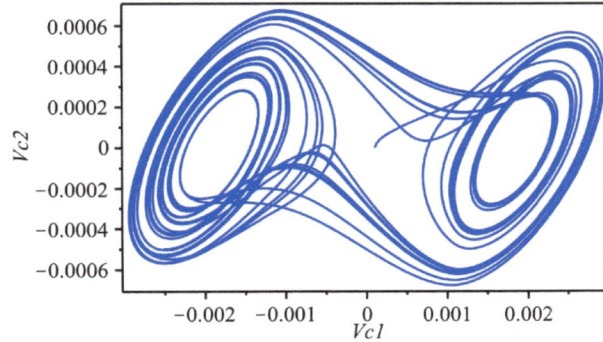

Figure 16: Double scroll attractor for Fig. **15**(b).

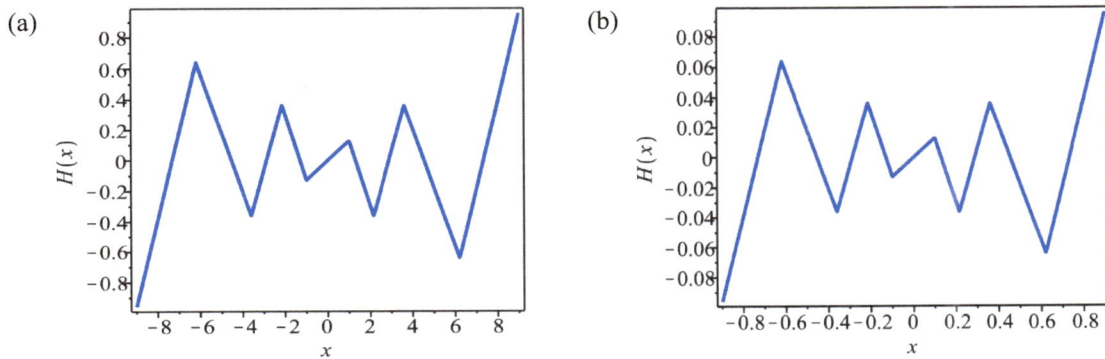

Figure 17: PWL function of generalized Chua's circuit: (a) without scaling and (b) with $\mu=10$.

Generalized Chua's Circuit

To scale the EL of the generalized Chua's circuit is necessary to add a parameter similar to Chua's circuit. By adding the parameter μ in the PWL function in (12), one modifies the breakpoints in the PWL description, consequently; (12) is recast by (48). The PWL function of Fig. **9**(b) and the PWL function with scaling factor of 10 are shown in Fig. **17**, respectively. The chaotic behaviour of the generalized Chua's circuit is not affected by this scaling of EL as shown in Fig. **18**.

$$h(x) = \left(m_{2q-1}\right)x + \frac{1}{2}\sum_{i=1}^{2q-1}\left(m_{i-1} - m_i\right) \times \left(\left|x + \frac{c_i}{\mu}\right| - \left|x - \frac{c_i}{\mu}\right|\right) \tag{48}$$

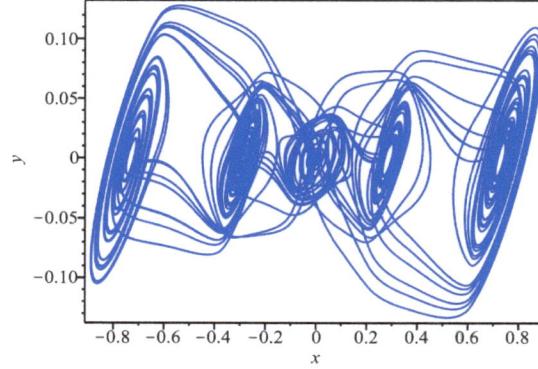

Figure 18: 5-scrolls attractor for Fig. **17**(b).

Multi-scroll Chaotic Oscillator

As one sees, figs. **10** to **14** show that the ELs are in a range of x(-30,30), y(-15,15) and x(-600,600), y(-400,400), respectively. Since real electronic devices cannot handle these ELs, (16) cannot be synthesized and it cannot have small ELs because k≥2 [79]. Consequently, h=2k or h=k for even or odd scrolls, respectively, to avoid superimposing of the slopes because the plateaus in (16) can disappear. Henceforth, the breakpoint of this saturated function (SF) is restricted to 1, so that to implement multi-scrolls attractors with multiple orientations using practical opamps one needs to scale ELs of the saturated function series. Then, the SF series in (16) is redefined by (49), where α allows that k<1 because the chaos-condition now applies on s=k/α, the new slope. In this manner, k and α can be selected to permit k<1, so that ELs in Fig. **10** to Fig. **14** can be scaled.

$$f(x;\alpha,k,h,p,q) = \begin{cases} (2q+1)k & x > qh + \alpha \\ k/\alpha(x - ih) + 2ik & |x - ih| \le \alpha, -p \le i \le q \\ (2i+1)k & ih + \alpha < x < (i+1)h - \alpha, -p \le i \le q-1 \\ -(2p+1)k & x < -ph - \alpha \end{cases} \tag{49}$$

As a result, 1D-3-scrolls attractor is now generated by setting a=b=c=d=0.7, k=250e-3, α=2.5e-3, s=100, h=250e-3, p=q=1 in (17) and (49), as shown in Fig. **19**; and 1D-6-scrolls attractor by setting a=b=c=d=0.7, k=250e-3, α=2.5e-3, s=100, h=500e-3, p=q=2 in (17) and (49), as shown in Fig. **20**.

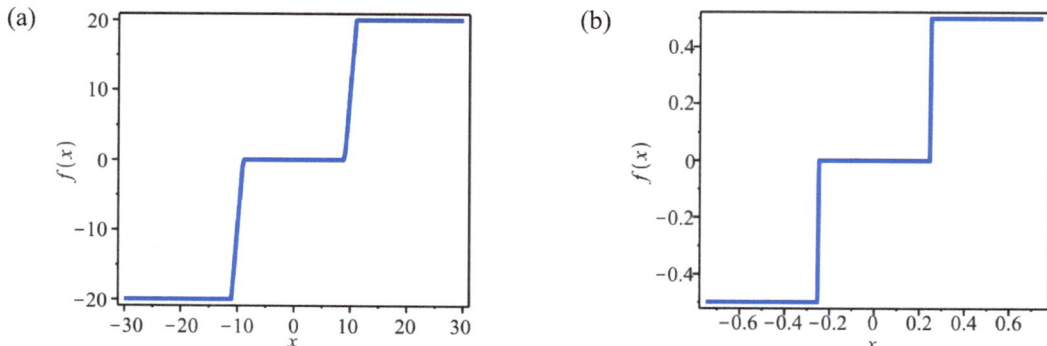

(c)

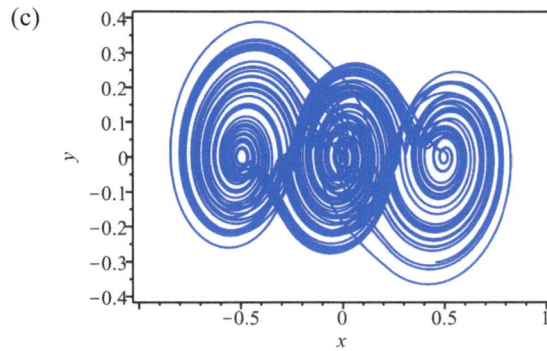

Figure 19: (a) SF without scaling for Fig. **10**(a), (b) SF with scaling and (c) its 1D-3-scrolls attractor.

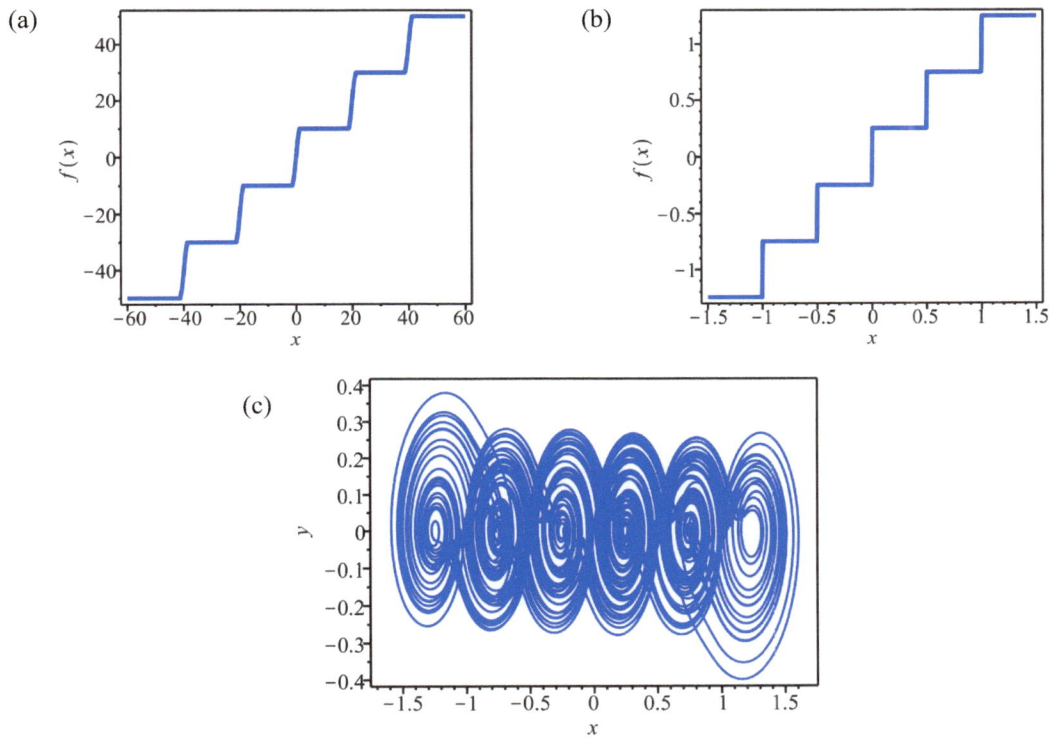

(a)

(b)

(c)

Figure 20: (a) SF without scaling for Fig. **10**(b), (b) SF with scaling and (c) its 1D-6-scrolls attractor.

Additionally, 2D-3-scrolls attractor is now generated by setting a=b=c=d_1=d_2=0.7, k1=k2=250e-3, α1=α2=2.5e-3, h1=h2=250e-3, p_1=q_1=p_2=q_2=1 in (18) and (49), as shown in Fig. **21**; and 2D-4-scrolls attractor is now generated with: a=b=c=d_1=d_2=0.7, k1=k2=250e-3, α1=α2=2.5e-3, h1=h2=500e-3, p_1=q_1=p_2=q_2=1 in (18) and (49), as shown in Fig. **22**. Furthermore, the position of scrolls on a 2D-mesh, the centers of scrolls and the connections among-neighbors scrolls are given by evaluating (19) to (23). For 2D-4-scrolls, it results in 16 scrolls with a radius of 250e-3 and 24 connections as shown in (50a). Similarly, the evaluation for 2D-3-scrolls is given in (50b) and it results in 9 scrolls with a radius of 250e-3 and 12 connections.

$$\mathbf{C} = \begin{bmatrix} (\pm 0.25, \pm 0.25) & (\pm 0.75, \pm 0.25) \\ (\pm 0.25, \pm 0.75) & (\pm 0.75, \pm 0.75) \end{bmatrix}$$

$$\mathbf{U}_x = \begin{bmatrix} (\pm 0.25, \pm 0.5) & (\pm 0.75, \pm 0.5) \end{bmatrix} \quad \mathbf{U}_y = \begin{bmatrix} (\pm 0.5, \pm 0.25) & (\pm 0.5, \pm 0.75) \end{bmatrix} \tag{50a}$$

$$\mathbf{U}_y = \begin{bmatrix} (\pm 0.5, \pm 0.25) & (\pm 0.5, \pm 0.75) \end{bmatrix} \quad \mathbf{U}' = \begin{bmatrix} (0, \pm 0.25) & (0, \pm 0.75) \end{bmatrix}$$

$$\mathbf{C}' = \begin{bmatrix} (0,0) & (\pm 0.5, 0) \\ (0, \pm 0.5) & (\pm 0.5, \pm 0.5) \end{bmatrix}$$

$$\mathbf{U}'_x = \left[(\pm 0.5, \pm 0.25) \right] \qquad \mathbf{U}'_y = \left[(\pm 0.25, \pm 0.5) \right]$$

$$\mathbf{U} = \left[(\pm 0.25, 0) \right] \qquad \mathbf{U}' = \left[(0, \pm 0.25) \right]$$

(50b)

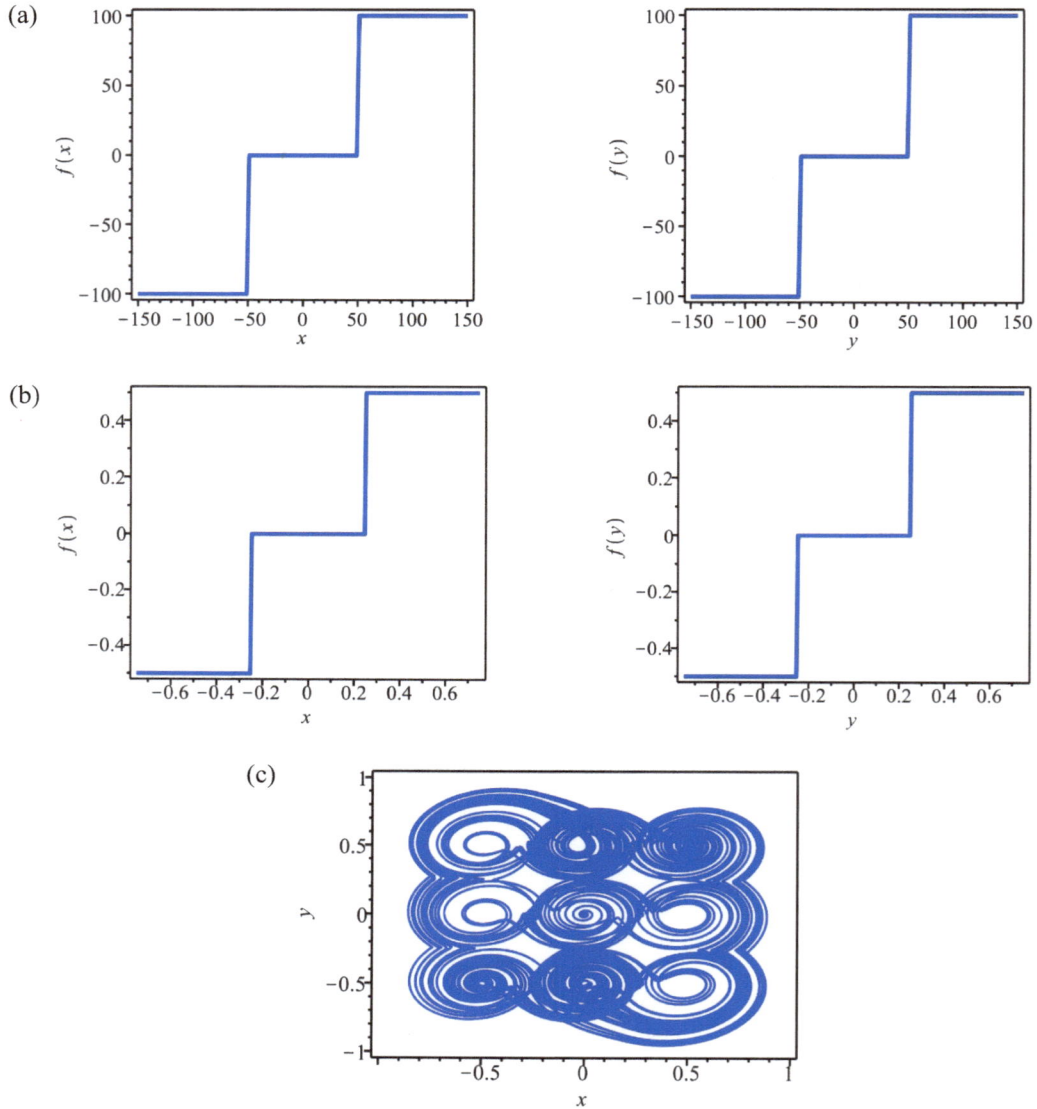

Figure 21: (a) SFx and SFy without scaling for Fig. **11**, (b) SFx and SFy with scaling and (c) its 2D-3-scrolls attractor.

Finally, 3D-3-scrolls attractor is now generated by setting $a = d_1 = 0.7$, $b = c = d_2 = d_3 = 0.8$, $k1 = 1$, $\alpha 1 = 10e\text{-}3$, $k2 = k3 = 0.5$, $\alpha 2 = \alpha 3 = 5e\text{-}3$, $h1 = 1$, $h2 = h3 = 0.5$, $p_1 = q_1 = p_2 = q_2 = p_3 = q_3 = 1$ in (24) and (49), as shown in Fig. **23**. 3D-4-scrolls attractor is now generated with: $a = d_1 = 0.7$, $b = c = d_2 = d_3 = 0.8$, $k1 = 1$, $\alpha 1 = 10e\text{-}3$, $k2 = k3 = 0.5$, $\alpha 2 = \alpha 3 = 5e\text{-}3$, $h1 = 2$, $h2 = h3 = 1$, $p_1 = q_1 = p_2 = q_2 = p_3 = q_3 = 1$ in (24) and (49), as shown in Fig. **24**. Now, the ELs of the attractors are within the ELs of real opamps.

Similarly, the position of scrolls on a 3D-mesh, the centers of scrolls and the connections among-neighbors scrolls can be also evaluated by (19) to (23). Besides, it is possible to have smaller ELs depending on the values of k and α.

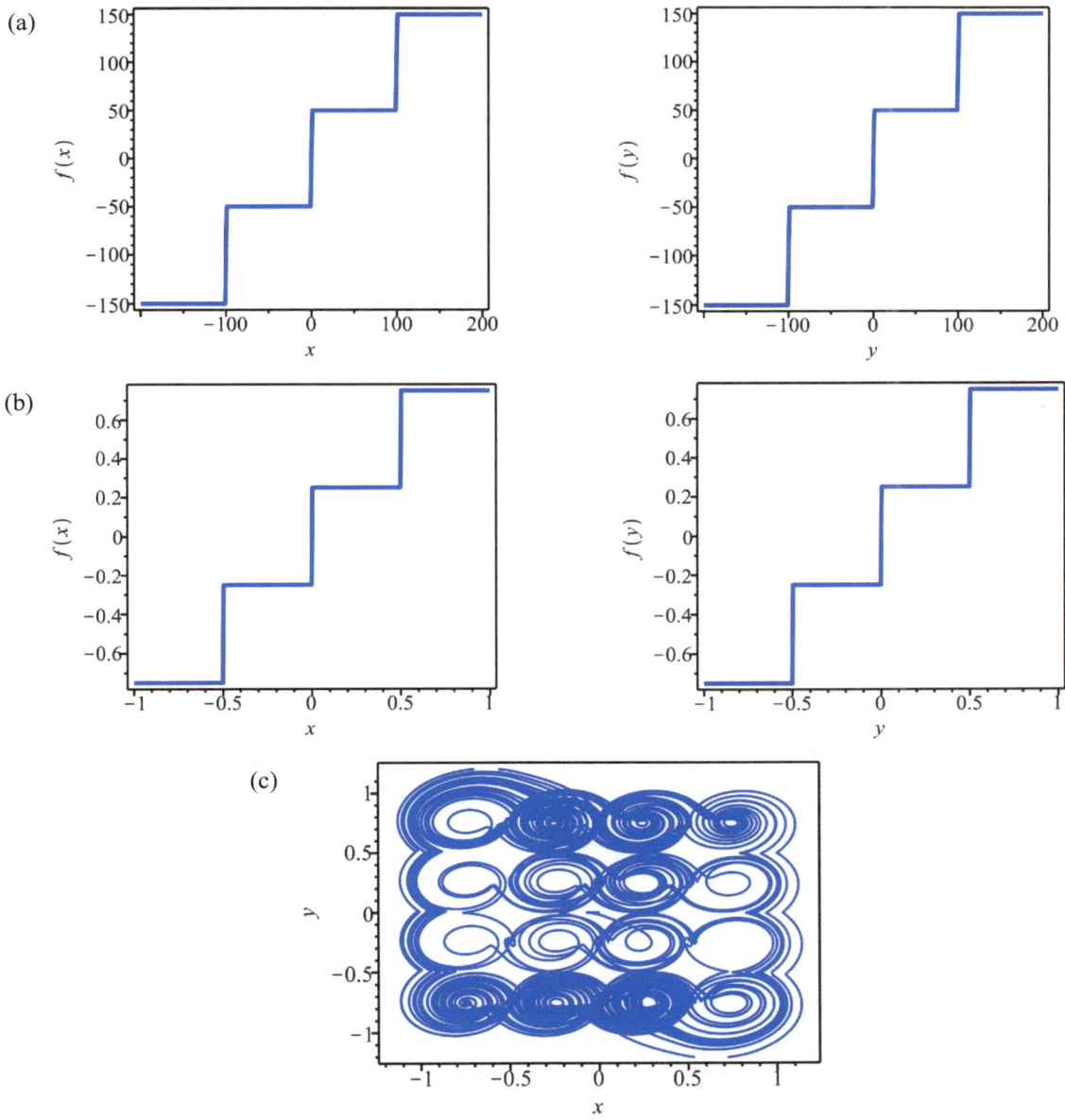

Figure 22: (a) SFx and SFy without scaling for Fig. **12**, (b) SFx and SFy with scaling and (c) its 2D-4-scrolls attractor.

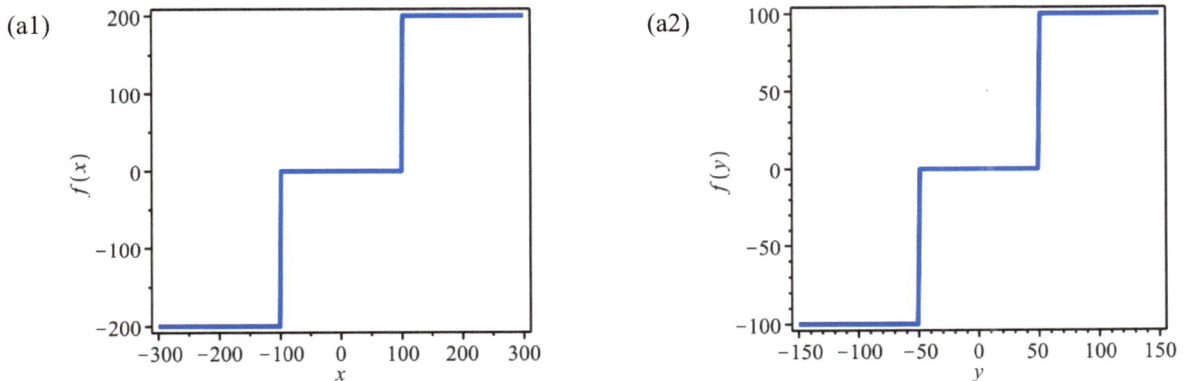

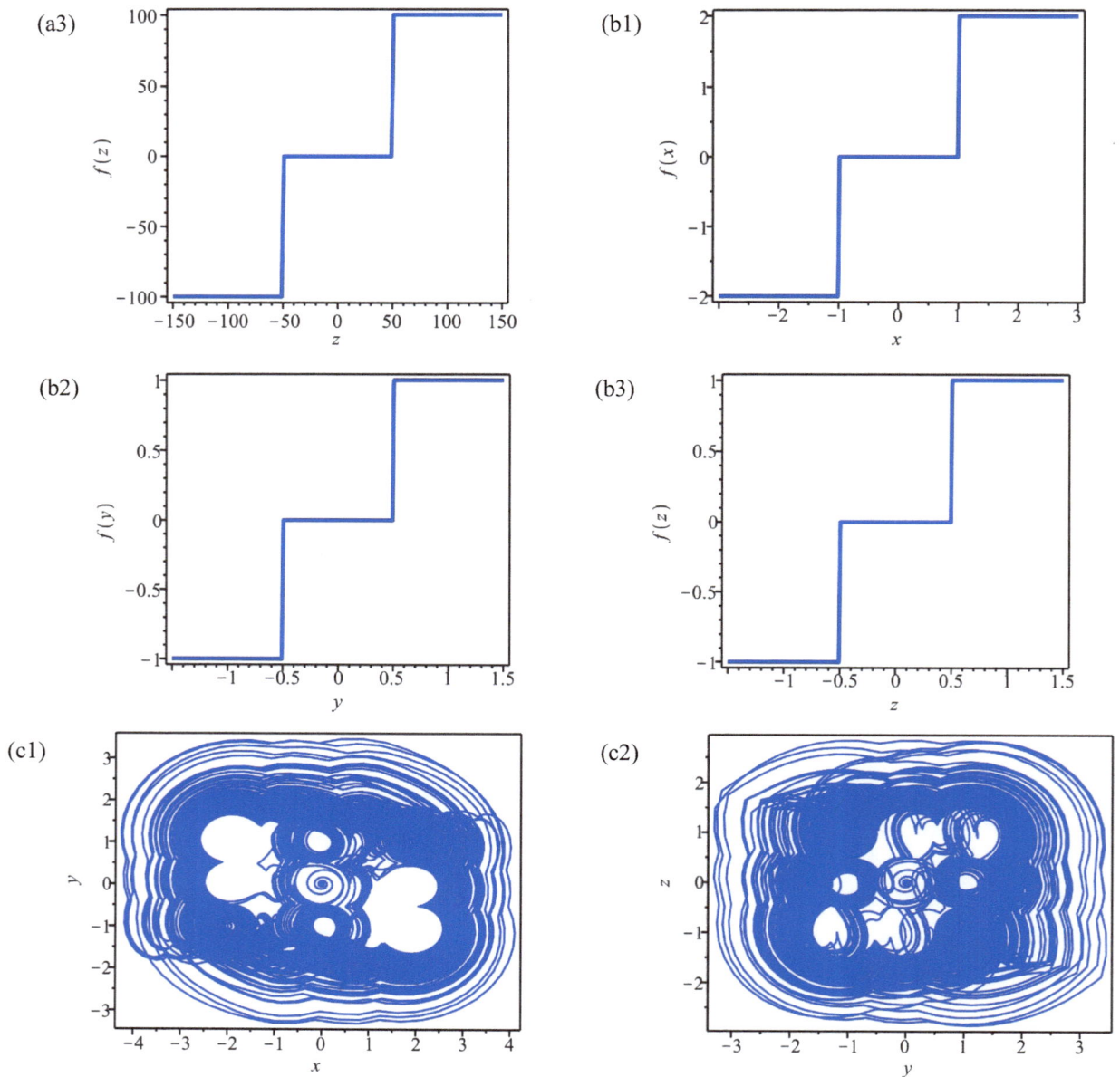

Figure 23: (a1), (a2), (a3) SFx, SFy and SFz without scaling for Fig. **13**; (b1), (b2), (b3) SFx, SFy and SFz with scaling and (c1), (c2) its 3D-3-scrolls attractor in x-y plane and y-z plane.

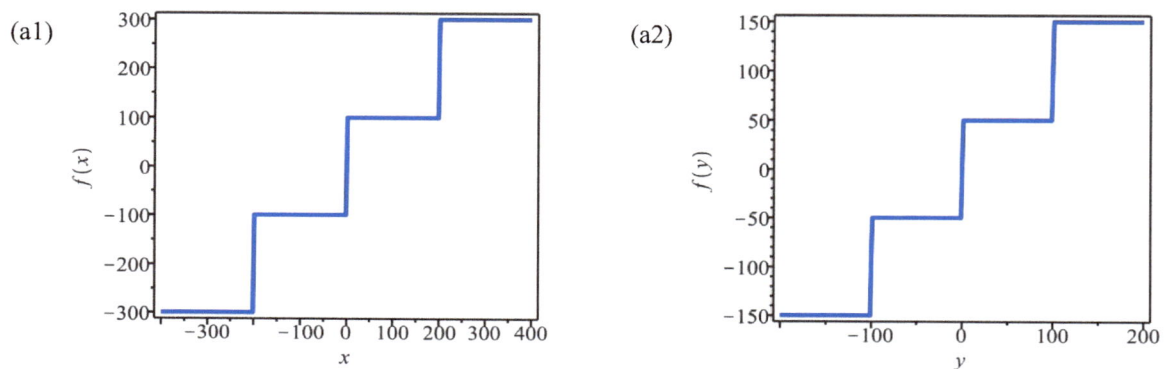

(a3)
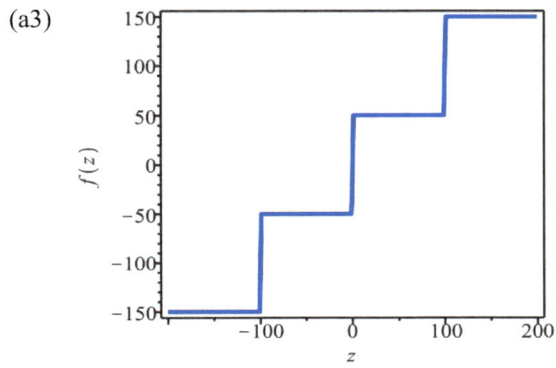

(b1)
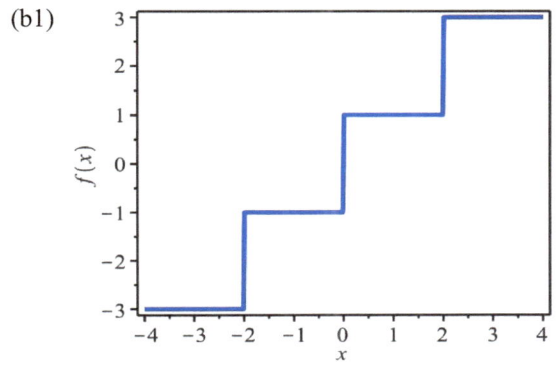

(b2)
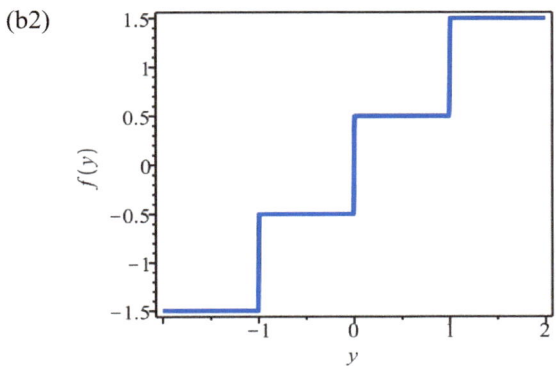

(b3)
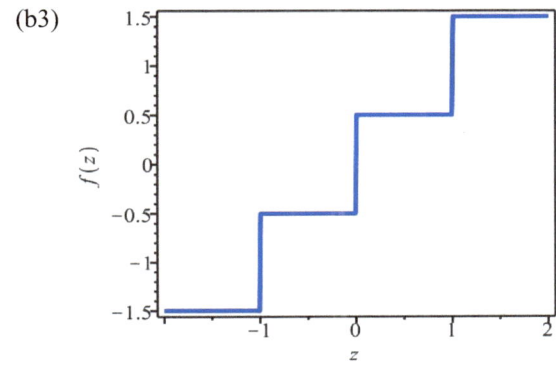

(c1)
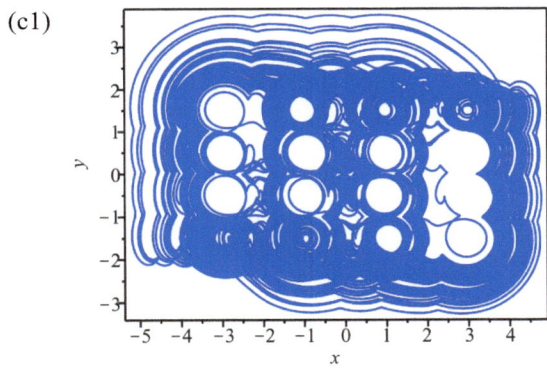

(c2)
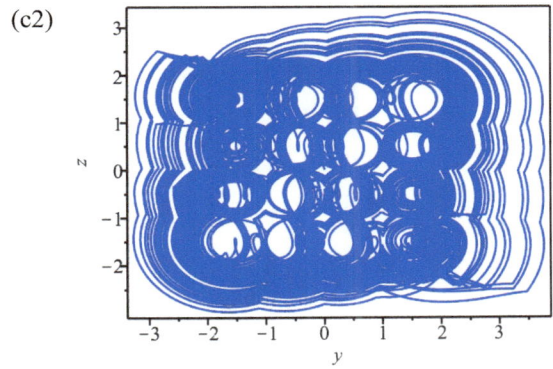

(d)
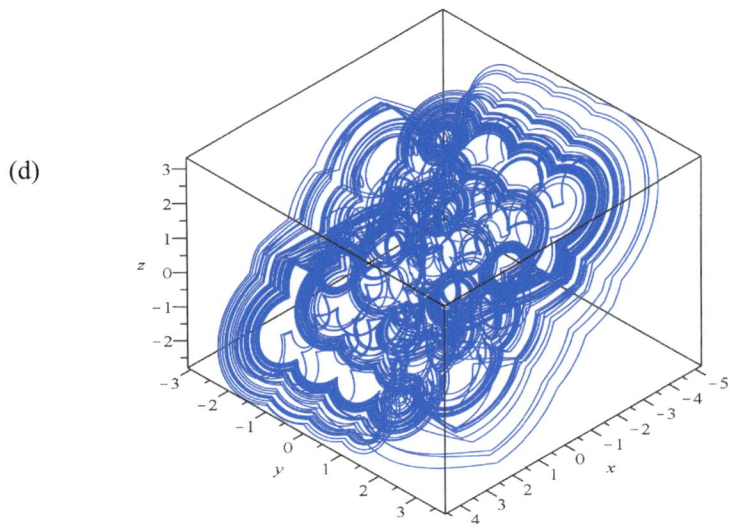

Figure 24: (a1), (a2), (a3) SFx, SFy and SFz without scaling for Fig. **14**; (b1), (b2), (b3) SFx, SFy and SFz with scaling; (c1), (c2) its 3D-4-scrolls attractor in x-y plane and y-z plane and (d) its three-dimensional representation.

FREQUENCY ESCALING OF CHAOTIC SIGNALS

The frequency scaling consists in selecting the frequency operation for the chaotic attractors. For synthesis purposes, this procedure is very useful because the designer can explore the frequency scaling by applying high-level simulations to synthesize multi-scrolls chaotic attractors using real opamps to implement the state variables systems and PWL functions. It should be pointed out that the frequency scaling is only limited by the finite bandwidth of commercial opamps [107,108].

There are two ways for implementing this procedure [130]. The first is carried out by multiplying the state variables systems by a required factor of scaling (FS) given by (51). The second is done on capacitors and/or inductors for the circuit implementations of the chaotic oscillators to evaluate new values of these elements given by (52). Therefore, one can select the frequency operation for chaotic attractors at highest level of abstraction using (51) and can further change this value of frequency at opamp-level of abstraction by using (52). The next subsections demonstrate the usefulness of the high-level simulations to estimate the values of the capacitors and inductors.

$$\dot{\mathbf{x}} = FS(\mathbf{Ax} + \mathbf{Bu}) \tag{51}$$

$$C_{fs} = \frac{C}{FS} \quad L_{fs} = \frac{L}{FS} \tag{52}$$

Chua's Circuit

In Fig. **25** is shown the signal of Vc1 in the time domain for Chua's circuit with a scaling frequency of FS=4 in (52) over the values of C1, C2 and L in Fig. **8**. Thereby, one obtains: R=1625, C1=112.5pF, C2=375pF, L=250µH, g1=1/1358, g2=1/2464, g3=1/1600, BP1=0.114V, BP2=0.4V. This result is compared with signal of Vc1 in Fig. **8**(a).

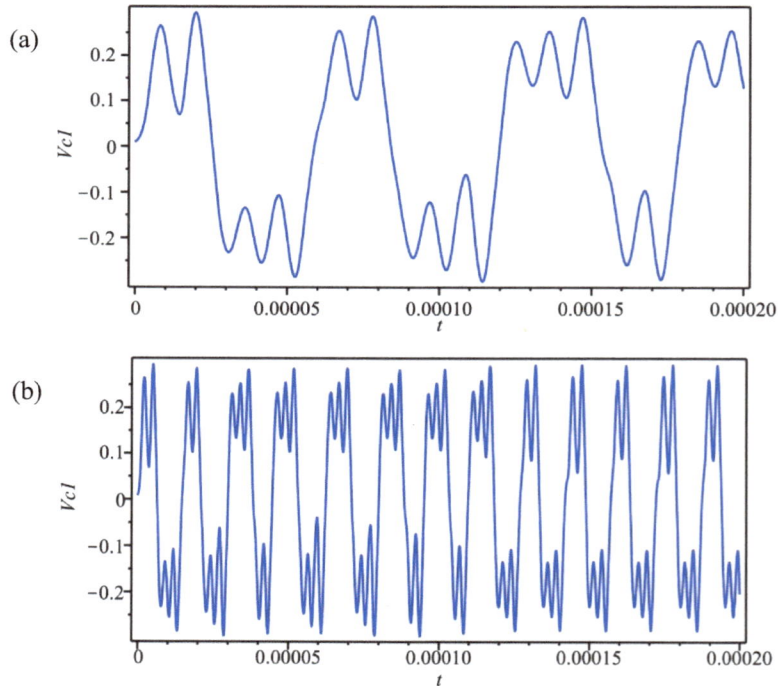

Figure 25: Signal of *Vc1* in time domain: (a) Fig **8**(a) and (b) its frequency scaling of 4.

Generalized Chua's Circuit

In Fig. **26** is shown the signal x in the time domain for generalized Chua's circuit with a scaling frequency of FS=3 in (51) over the state variables system in (14). Thereby, by selecting $q = 3$, $m = [0.9/7, -3/7, 3.5/7, -2.7/7, 4/7, -2.4/7]$, $c = [1; 2.15; 3.6; 6.2; 9]$, with $\alpha = 9$, $\beta = 14.28$ $\delta = 1$ in (12) and (14), respectively, one obtains (53). This result is compared with signal x in Fig. **18**.

$$\begin{bmatrix} \dot{x} \\ \dot{y} \\ \dot{z} \end{bmatrix} = \begin{bmatrix} -3\alpha(1+\delta) & 3\alpha & 0 \\ 3 & -3 & 3 \\ 0 & -3\beta & 0 \end{bmatrix} \begin{bmatrix} x \\ y \\ z \end{bmatrix} + \begin{bmatrix} 3\alpha f(x) \\ 0 \\ 0 \end{bmatrix} \tag{53}$$

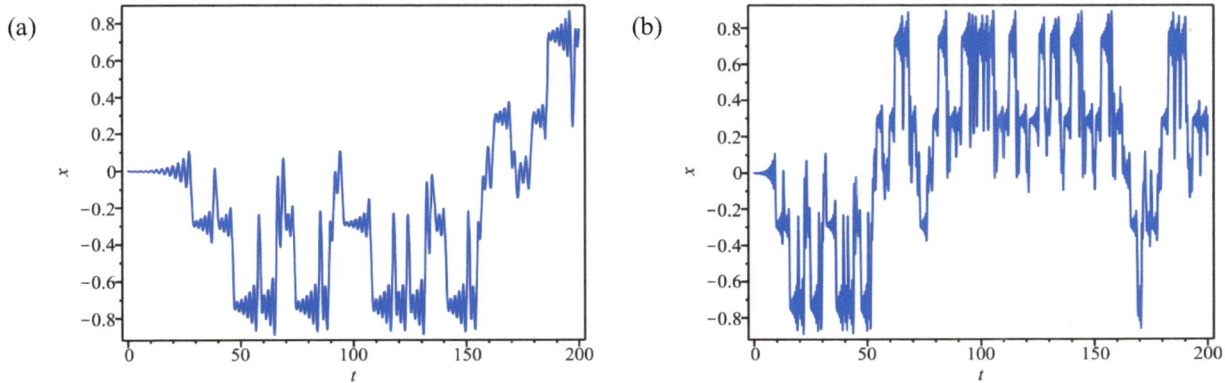

Figure 26: Signal x in time domain: (a) Fig 3.18 and (b) its frequency scaling of 3.

Multi-scrolls Chaotic Oscillator

The frequency scaling is only applied to 3D-4-scroll chaotic attractor for sake of simplicity; however, it can be applied in the same way to others (1D and 2D-multi-scrolls attractors). In Fig. **27** is shown the signal x in the time domain for 3D-4-scroll chaotic attractor with a scaling frequency of FS=5 in (51) over the state variables system in (24). Thereby, by selecting a=d_1=0.7, b=c=d_2=d_3=0.8, k1=1, α1=10e-3, k2=k3=0.5, α2=α3=5e-3, h1=2, h2=h3=1, p_1=q_1=p_2=q_2= p_3=q_3=1 in (24) and (49), respectively, one obtains (54). This result is compared with signal x in Fig. **24**(c).

$$\dot{x} = 5y - 5\frac{d_2}{b} f(y; k_2, h_2, p_2, q_2) \qquad \dot{y} = 5z - 5\frac{d_3}{c} f(z; k_3, h_3, p_3, q_3)$$

$$\dot{z} = -5ax - 5by - 5cz + 5d_1 f(x; k_1, h_1, p_1, q_1) + 5d_2 f(y; k_2, h_2, p_2, q_2) + 5d_3 f(z; k_3, h_3, p_3, q_3) \tag{54}$$

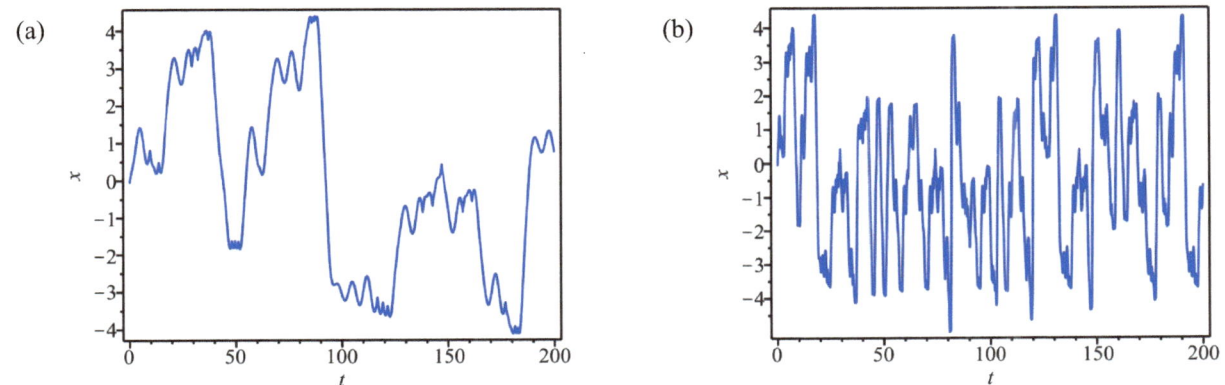

Figure 27: Signal x of 3D-4 scrolls in time domain: (a) Fig 3.24(c), and (b) its frequency scaling of 5.

LYAPUNOV EXPONENTS

The deterministic, still unpredictable behaviour of nonlinear dissipative dynamical systems is an important subject in more and more fields of science, from mathematics to biology; even in engineering. The main characterizations of chaotic systems are fractal dimension, Kolmogorov-Sinai entropy and Lyapunov spectrum [131]. Their relationship is established through the Kaplan-Yorke conjecture [132]. Among them, the Lyapunov exponents give the most characteristic description of the presence of a deterministic non-periodic flow [133]. Therefore, Lyapunov exponents are asymptotic measures characterizing the average rate of growth (or shrinking) of small perturbations to the solutions of a dynamical system. Lyapunov exponents provide quantitative measures of response sensitivity of a dynamical system to small changes in initial conditions [131]. One feature of chaos is the sensitive dependence on initial conditions, for non-chaotic systems all the Lyapunov exponents are non-positive. Therefore, the presence of positive Lyapunov exponents has often been taken as a signature of chaotic motion [131-133].

In this manner, an algorithm capable of computing the Lypunov exponents in a simple fashion for chaos synthesis approaches is required. From this point of view, a simple algorithm to compute the Lyapunov exponents of multi-scrolls chaotic systems is introduced herein. The idea is to trace the evolution of a set of vectors where the orthogonality among these vectors is kept by using the Gram-Schmidt method. The key difference relapses in that a nonlinear system and its associated variational equation is transformed on *m*-piecewise linear (PWL) systems and their associated *m*-piecewise (PW) variational equations, which are constants. Consequently, to speed-up time simulation, a solution of these *m*-PW variational systems is computed by applying the Forward-Euler algorithm for each time step size h. Due to this, the algorithm could be implemented into the workflow of automated synthesis tools where several solutions must be evaluated in order to determine if the potential solution presents chaotic behaviour.

Computation of Lyapunov Exponents

For means of illustration, consider an *n*-dimensional dynamical system

$$\dot{\mathbf{x}} = \mathbf{f}(\mathbf{x}) \quad t \geq 0 \quad \mathbf{x}(0) = \mathbf{x}_0 \in R^n \tag{55}$$

where $\mathbf{x}$ $\mathbf{y}$ $\mathbf{f}$ are *n*-dimensional vector fields. To determine the *n*-Lyapunov exponents of the system we have to find the long term evolution of small perturbations to a trajectory, which are determined by the variational equation of (55) [132],

$$\dot{\mathbf{y}} = \frac{\partial \mathbf{f}}{\partial \mathbf{x}}(\mathbf{x}(t))\mathbf{y} = \mathbf{J}(\mathbf{x}(t))\mathbf{y} \tag{56}$$

where $\mathbf{J}$ is the $n \times n$ Jacobian matrix of $\mathbf{f}$. A solution of (56) with a given initial perturbation $\mathbf{y}(0)$ can be written as

$$\mathbf{y}(t) = \mathbf{Y}(t)\mathbf{y}(0) \tag{57}$$

with $\mathbf{Y}(t)$ as the fundamental solution satisfying

$$\dot{\mathbf{Y}} = \mathbf{J}(x(t))\mathbf{Y} \quad \mathbf{Y}(0) = \mathbf{I}_n \tag{58}$$

Here $\mathbf{I}_n$ denotes the $n \times n$ identity matrix. If we consider the evolution of an infinitesimal *n*-parallelepiped $[p_1(t),...,p_n(t)]$ with the axis $\mathbf{p}_i(t) = \mathbf{Y}(t)\mathbf{p}_i(0)$ for $i = 1,...,n$, where $\mathbf{p}_i(0)$ denotes an orthogonal basis of R^n. The *ith* Lyapunov exponent, which measures the long-time sensitivity of the flow $\mathbf{x}(t)$ with respect to the initial data $\mathbf{x}(0)$ at the direction $\mathbf{p}_i(t)$, is defined by the expansion rate of the length of the *ith* axis $\mathbf{p}_i(t)$ [133] and is given by

$$\lambda_i = \lim_{t \to \infty} \frac{1}{t} \ln \|\mathbf{p}_i(t)\| \tag{59}$$

Therefore, it is discussed a new and simple method based on the solution of (58).

Simple Method for Computing Lyapunov Exponents

By considering that a chaotic system described by (55) can be modelled by applying state variables approach, where the nonlinear term of chaotic systems is represented by PWL characteristics as described in section 3.1. Recall that, the general form for these PWL descriptions is $f(x_j) = r_{ij}x_j + b_{ij}$, being r and b the slope and the offset for each one of the segments in the PWL characteristic, respectively. The main idea behind is to transform the problem of solving a nonlinear system of differential equations into a sequence of linear and purely algebraic problem which can be solved straightforward by applying a simple numerical integration algorithm. Consequently, the PWL characteristic can be divided in two matrices given by

$$\mathbf{A}' = \begin{bmatrix} r_{11} & \cdots & r_{1j} \\ \vdots & \ddots & \vdots \\ r_{i1} & \cdots & r_{ij} \end{bmatrix} \quad \mathbf{B}' = \begin{bmatrix} b_{11} & \cdots & b_{1j} \\ \vdots & \ddots & \vdots \\ b_{i1} & \cdots & b_{ij} \end{bmatrix} \tag{60}$$

here, i and j denotes the position where the slope and offset must be placed according to the state variables. Therefore, (55) is recast by (61) and (62), which generate a set of systems of first-order linear differential equations (*m*-PWL systems) according to number of regions.

$$\dot{\mathbf{x}} = [(\mathbf{A} + \mathbf{A}')\mathbf{x}] + (\mathbf{B} + \mathbf{B}')\mathbf{u} \tag{61}$$

$$\dot{\mathbf{x}} = \hat{\mathbf{A}}_m\mathbf{x} + \hat{\mathbf{B}}_m\mathbf{u} \tag{62}$$

where $\hat{\mathbf{A}}$ and $\hat{\mathbf{B}}$ are matrices of appropriate dimensions, $\mathbf{u}$ is the input vector and m indicates the number of regions. Note that the x-dependent part in (62) is defined only by $\dot{\mathbf{x}} = \hat{\mathbf{A}}_m\mathbf{x}$. This gives a solution of (58) in the form of

$$\dot{\mathbf{Y}}_m = \mathbf{J}_m(x)\mathbf{Y}_m \quad \mathbf{Y}_m(0) = \mathbf{I}_n \tag{63}$$

yielding the *m*-PW variational systems where $\mathbf{J}_m(x) = \partial\hat{\mathbf{A}}_m\mathbf{x}/\partial\mathbf{x}$. Here, the *m*-Jacobian matrices are constants.

In this manner, if one chooses an orthogonal basis denoted by $\mathbf{Q}_k = [\mathbf{q}_{1,k}, \mathbf{q}_{2,k}, \ldots, \mathbf{q}_{n,k}]$, with the vectors $\mathbf{q}$ being the Lyapunov vectors at time t_k; the method proceeds to integrate (63) for obtaining $\mathbf{Y}_{m(k+1)}$ with initial condition $\mathbf{Y}_m(t_k) = \mathbf{Q}_k$. Furthermore, the Gram-Schmidt orthogonalization [131], given by (64) is applied iteratively to the matrix $\mathbf{Y}_{m(k+1)}$ for obtaining $\mathbf{Q}_{k+1}$.

$$\mathbf{q}_{(1,k+1)} = \frac{\mathbf{y}_{m(1,k+1)}}{\left\| \mathbf{y}_{m(1,k+1)} \right\|} \quad \mathbf{q}_{(i,k+1)} = \frac{\mathbf{y}_{m(i,k+1)} - \sum_{ii=1}^{i-1}\left(\mathbf{y}_{m(i,k+1)}\cdot\mathbf{q}_{(ii,k+1)}\right)\mathbf{q}_{(ii,k+1)}}{\left\| \mathbf{y}_{m(i,k+1)} - \sum_{ii=1}^{i-1}\left(\mathbf{y}_{m(i,k+1)}\cdot\mathbf{q}_{(ii,k+1)}\right)\mathbf{q}_{(ii,k+1)} \right\|} \quad i = 1,\ldots,n \tag{64}$$

Finally, the Lyapunov exponents are accumulated from the norm of the vectors $\mathbf{q}_{(1,k+1)}$ and $\mathbf{q}_{(i,k+1)}$ in the form of

$$\lambda_i = \lim_{k\to\infty}\frac{1}{t_{k+1}}\sum_{L=1}^{k+1}\ln\left\| \mathbf{y}_{m(i,L)} - \sum_{ii=1}^{i-1}(\mathbf{y}_{m(i,L)}\cdot\mathbf{q}_{(ii,L)})\mathbf{q}_{(ii,L)} \right\| \tag{65}$$

Note that only one variational system $\dot{\mathbf{Y}}_m$ is integrated in each step size h and it depends on the region in the PWL characteristic. As mentioned above, the *m*-Jacobian matrices are constants so that the numerical time integration used to discretize (62) and (63) is based on the Forward-Euler method.

Renormalization of the Lyapunov Exponents

As shown in section 3.7, the frequency scaling approach can be applied to chaotic systems. However, if this scaling is not considered, the magnitude of the corresponding Lyapunov exponent will grow accordingly the frequency scaling applied and it leads to incorrect values for the Lyapunov exponents. The problem can be easily resolved by normalizing the Lyapunov exponents at every time step. Renormalization will not affect the orthogonality among the **q** vectors. The renormalization factor is obtained from $RNF = 1/\lambda_s$ by calculating the smallest absolute value of all the eigenvalues of m-Jacobian matrices in (63). This originates that (65) is recast by (66)

$$\lambda_i = \lim_{k\to\infty}\left[\frac{1}{t_{k+1}}\sum_{L=1}^{k+1}\ln\left\|\mathbf{y}_{m(i,L)} - \sum_{ii=1}^{i-1}(\mathbf{y}_{m(i,L)}\cdot\mathbf{q}_{(ii,L)})\mathbf{q}_{(ii,L)}\right\|\right]*RNF \tag{66}$$

Error in Orthogonality

The algorithm previously introduced depends on the Gram-Schmidt orthogonalization, which is iteratively applied to the matrix $\mathbf{Y}_{m(k+1)}$ in (63). To check for the accuracy of the algorithm, it will be convenient to consider the error in orthogonality [132]. This is calculated by applying (67).

$$error = \left\|\mathbf{Y}_{m(k+1)}\mathbf{Y}_{m(k+1)}^{\mathrm{T}} - \mathbf{I}\right\| \tag{67}$$

where $\mathbf{Y}_{m(k+1)}$ and $\mathbf{Y}_{m(k+1)}^{\mathrm{T}}$ are the matrices generated by the Gram-Schmidt process in t_{k+1} and its transpose, respectively. To ensure small errors for the approximations described in (62) and (63), a small step size h needs to be used to compute the dynamical solution $x(t)$ of (55).

Chua`s Circuit

By applying the proposed algorithm to Chua`s circuit, one obtains (68).

$$m=1\quad\begin{aligned}\mathbf{x}_{k+1} &= \mathbf{x}_k + h(\hat{\mathbf{A}}_1\mathbf{x}_k + \hat{\mathbf{B}}_1)\\ \mathbf{Y}_{1(k+1)} &= \mathbf{Q}_k + h(\mathbf{J}_1\mathbf{Q}_k)\end{aligned}\quad x_1 < -BP1$$

$$m=2\quad\begin{aligned}\mathbf{x}_{k+1} &= \mathbf{x}_k + h(\hat{\mathbf{A}}_2\mathbf{x}_k + \hat{\mathbf{B}}_2)\\ \mathbf{Y}_{2(k+1)} &= \mathbf{Q}_k + h(\mathbf{J}_2\mathbf{Q}_k)\end{aligned}\quad -BP1 \le x_1 \le BP1 \tag{68}$$

$$m=3\quad\begin{aligned}\mathbf{x}_{k+1} &= \mathbf{x}_k + h(\hat{\mathbf{A}}_3\mathbf{x}_k + \hat{\mathbf{B}}_3)\\ \mathbf{Y}_{3(k+1)} &= \mathbf{Q}_k + h(\mathbf{J}_3\mathbf{Q}_k)\end{aligned}\quad x_1 > BP1$$

By solving (68) and using (66) with RNF=0.000001, the spectrum of Lyapunov exponents for double-scroll attractor in Fig. **8** are obtained as shown in Fig. **28**. Their sum and error are shown in Fig. **29**. Thereby, the Lyapunov exponents are LE1=0.357, LE2=0.010 and LE3=-5.998. Besides, one can evaluate the Lyapunov dimension [132] according to (69), which is given by LD=2.062.

$$D_L = j + \frac{1}{\left|LE_{j+1}\right|}\sum_{i=1}^{j}LE_i \tag{69}$$

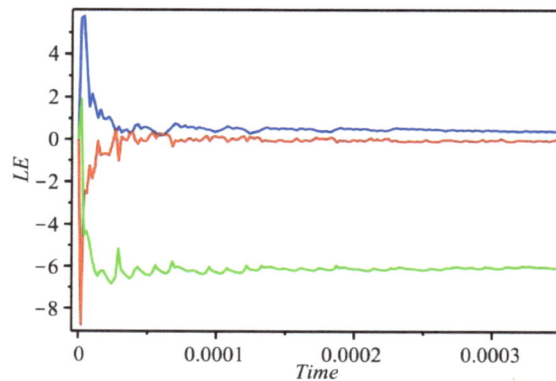

Figure 28: Evolution of the Lyapunov exponents of Chua's circuit.

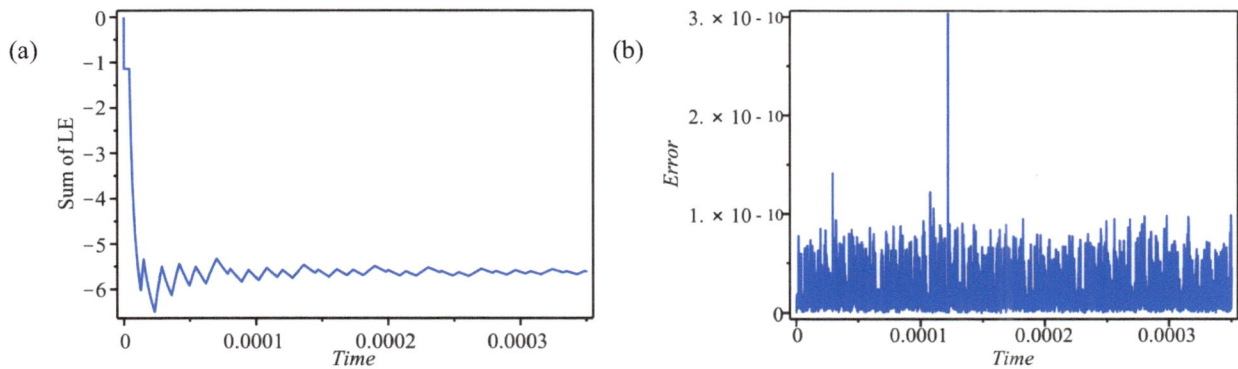

Figure 29: (a) Sum of the Lyapunov exponents and (b) its error of Fig. 28.

Multi-scrolls Chaotic Oscillator

According to the above analysis, the Lyapunov exponents for the 1D-multi-scrolls chaotic system defined by (17) and (49) are calculated by applying the proposed algorithm. In this manner, one obtains LE1=0.121, LE2=0.006, LE3=-0.77 and LD=2.165 with RNF=0.00001 as shown in Fig. **30**, and their sum and error are shown in Fig. **31**.

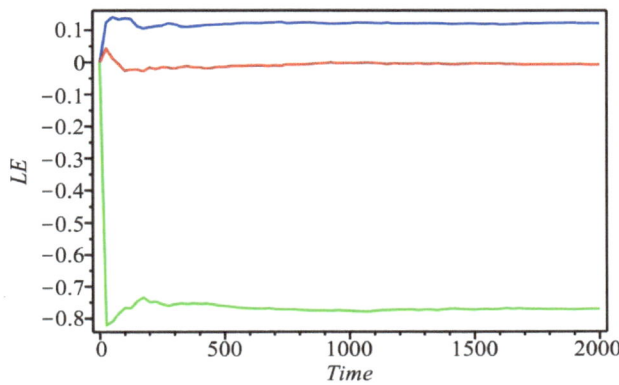

Figure 30: Evolution of the Lyapunov exponents for 1D-6-scrolls chaotic attractor in Fig. 20(c).

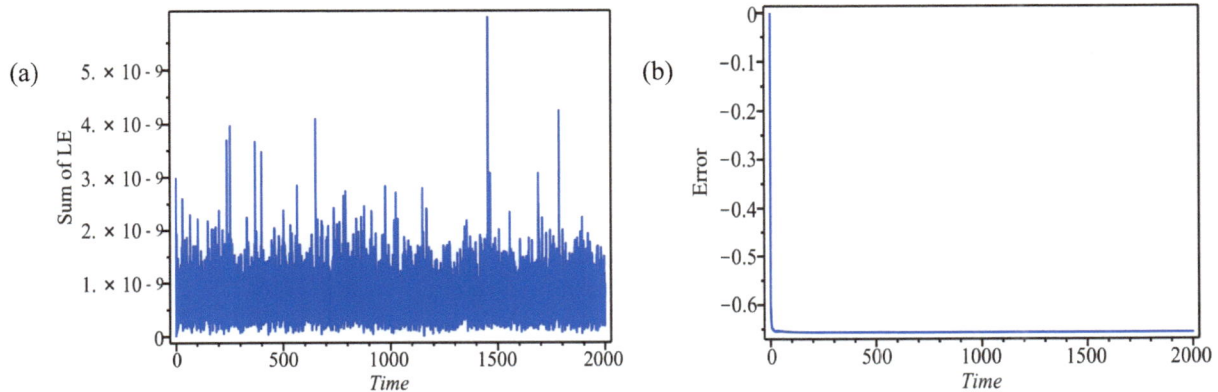

Figure 31: (a) Sum of the Lyapunov exponents and (b) its error of Fig. 30.

In the same way, one can compute the Lyapunov exponents for 2D and 3D multi-scrolls chaotic attractors. In Fig. **32** and **33** are shown the Lyapunov exponents, their sum and error, respectively. By applying the proposed algorithm, one obtains LE1=0.141, LE2=0.002, LE3=-0.803 and LD=2.178 with RNF=0.00001.

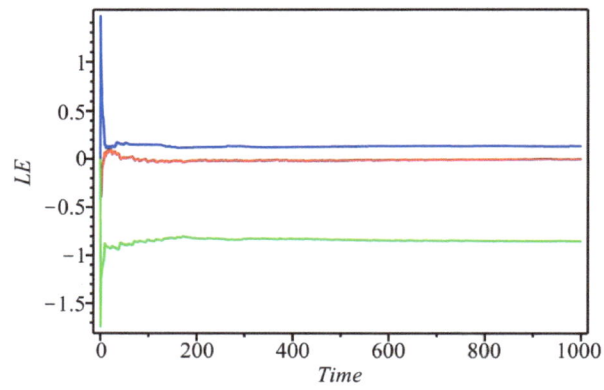

Figure 32: Evolution of the Lyapunov exponents for 2D 4-scrolls chaotic attractor in Fig. 22(c).

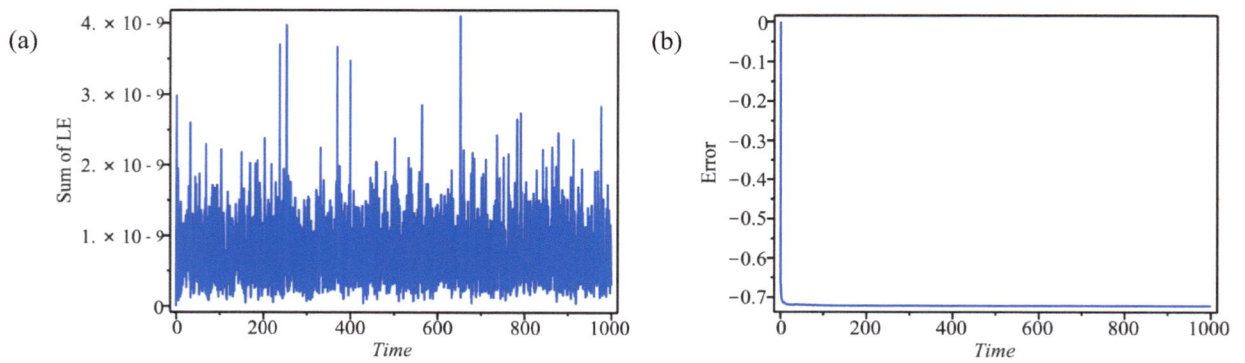

Figure 33: (a) Sum of the Lyapunov exponents and (b) its error of Fig. 32.

CHAPTER 4

Synthesis of PWL Functions

Abstract: The design of a nonlinear function with multi-segments is a basis for generating chaotic attractors with multidirectional orientation and with a large number of scrolls. However, it has been identified that it is quite difficult to synthesize nonlinear functions with multi-segments by using analog electronic circuits. Therefore, to cope with this problem, the electronic design automation (EDA) industry is developing design tools with a high degree of abstraction (behavioural modelling). Henceforth, the synthesis approach presented in this chapter depends on the behavioural modelling of the chaotic systems introduced in Chapter. 3. In particular, this approach is focused on designing multi-scrolls chaotic attractors with multidirectional orientation. In this manner, this chapter first presents a basic cell based on opamps to synthesize saturated nonlinear functions series. Further, a new synthesis approach is introduced to design the saturated functions in current and voltage modes by using a general scheme of connection for basic cells. Finally, at the end of this chapter, a Verilog-A model for considering no-ideal effects of the opamps is defined to execute SPICE simulations.

Keywords: Chaos, electronic design automation, dynamical systems, chaos generators, modelling and simulation.

OPERATIONAL AMPLIFIER FINITE GAIN MODEL

This section reviews the saturated function concept and presents some fundamental conditions for generating multi-scrolls chaotic attractors using opamps. In Chapter 3, in section 3.1.3 it was introduced a chaotic system based on saturated nonlinear function series, which can generate multi-scrolls according to the number of segments in the function series. For instance, let's consider the saturated function series given in (1), where k>0 is the slope of the middle segment and is called the saturated slope, and the upper and lower radial are called saturated plateaus. Fig. **1** shows the characteristic of this saturated nonlinear function series.

$$f(x) = \begin{cases} k & x > 1 \\ kx \rightarrow |x| \leq 1 \\ -k & x < -1 \end{cases}$$

(1)

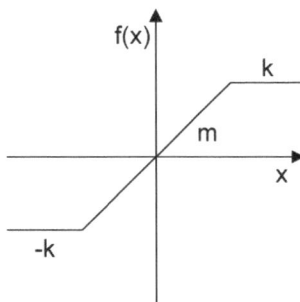

Figure 1: Saturated function series.

On the other hand, it is well known that a saturated circuit is one of the basic PWL circuits. The PWL models for opamps can be well characterized by saturated circuits. Fig. **2** shows that the PWL approximations using opamps are quite accurate [96]. It leads to the representation for the opamp given by (2), which is in the linear region for $-Vsat \leq v_o \leq Vsat$ with voltage amplification Av, positive saturation Vsat and negative saturation $-Vsat$ [3].

$$v_0 = \frac{Av}{2}\left(\left|vi + \frac{Vsat}{Av}\right| - \left|vi - \frac{Vsat}{Av}\right|\right) i(-) = i(+) = 0$$

(2)

Jesus Manuel Muñoz Pacheco and Esteban Tlelo Cuautle

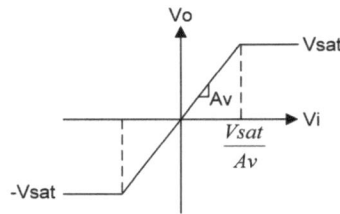

Figure 2: Finite gain model for opamps.

The DC transfer shown in Fig. **2** is called the opamp finite gain model [96]. In each of the three regions the opamp can be characterized by a linear circuit. As one sees, a saturated function series can be related to the opamp finite gain model as shown in the next section.

SATURATED FUNCTION AND ITS CIRCUIT IMPLEMENTATION

A saturated function working in voltage mode is described by (3), where ±A are the values of the saturated voltages, sp are the switching points, and (A/sp) is the slope. Fig. **3**(a) shows the voltage saturated function of (3). Similarly, a shifted voltage saturated function is described by (4) and (5), where ±A are the values of the saturated voltages, E_1 is the shifted voltage, $\pm E_1 \pm sp$ are the switching points, and (A/sp) is the slope. Fig. **3**(b) and **3**(c) shows the shifted voltage in the saturated functions. According to the value of E, one gets voltage saturated functions with positive (Fig. **3** (c)) and negative shifts (Fig. **3** (b)).

$$F_S(v) = \begin{cases} A & v > sp \\ (A/sp)v & -sp \leq v \leq sp \\ -A & v < -sp \end{cases} \tag{3}$$

$$F_S(v-E) = \begin{cases} A & v > sp + E_1 \\ (A/sp)v & -sp + E_1 \leq v \leq sp + E_1 \\ -A & v < -sp + E_1 \end{cases} \tag{4}$$

$$F_S(v+E) = \begin{cases} A & v > sp - E_1 \\ (A/sp)v & -sp - E_1 \leq v \leq sp - E_1 \\ -A & v < -sp - E_1 \end{cases} \tag{5}$$

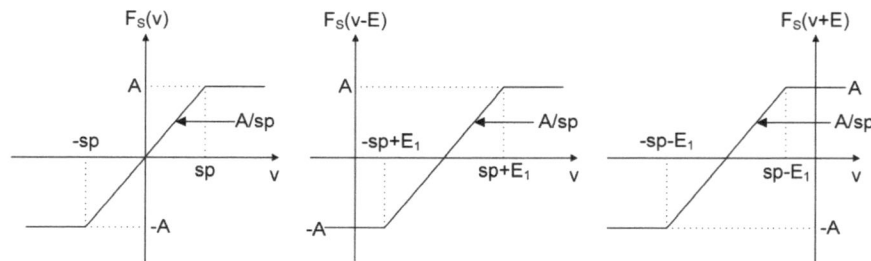

Figure 3: (a) Voltage saturated function, (b) with negative shift and (c) with positive shift.

Based on the voltage saturated functions in (3)-(5), one can further define the current saturated functions as given in (6), where R is a voltage-to-current conversion resistor. One can easily realize the transformation between the voltages saturated functions and the current saturated functions via the voltage-to-current conversion resistor.

$$F_S(i) = \frac{F_S(v \pm E)}{R} \tag{6}$$

Consequently, if one relates the opamp finite-gain model in (2) and shown in Fig. **2** with (**3-5**), one gets voltage saturated functions, which depend now on the opamp model [107]. These saturated functions series are given by (7) for no shift, positive and negative shifts, respectively.

$$Vo = \frac{Av}{2}\left(\left|Vi + \frac{Vsat}{Av}\right| - \left|Vi - \frac{Vsat}{Av}\right|\right)$$

$$Vo = \frac{Av}{2}\left(\left|Vi + \frac{Vsat}{Av} - E\right| - \left|Vi - \frac{Vsat}{Av} - E\right|\right)$$

$$Vo = \frac{Av}{2}\left(\left|Vi + \frac{Vsat}{Av} + E\right| - \left|Vi - \frac{Vsat}{Av} + E\right|\right)$$

$$(7)$$

Now, the parameters k, α, h, s of the PWL function to generate multi-scrolls chaotic attractors defined by (3.49), are determined by (7), where α=Vsat/Av are the breakpoints, k=Vsat is the saturated plateau, and s=Vsat/α is the saturated slope. To generate saturated functions as shown in Fig. **3**, E takes different values in (7) to synthesize the required plateaus and slopes. The opamp based-basic cell shown in Fig. **4** is herein used to synthesize current saturated functions from (7) via the voltage-to-current conversion resistor Rc. The values of the plateaus k, in voltage and current, the breakpoints α, the slope s and h are evaluated by (8) and depend on the circuit parameters of Fig. **4**.

$$k = RixIsat, \quad Isat = \frac{Vsat}{Rc}, \quad \alpha = \frac{Ri|Vsat|}{Rf}, \quad s = \frac{k}{\alpha}, \quad h = \frac{Ei}{\left(1 + \dfrac{Ri}{Rf}\right)}$$

$$(8)$$

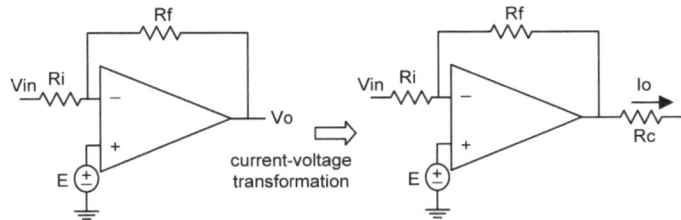

Figure 4: Opamp based-basic cell to generate the saturated functions of (7).

SYNTHESIS OF CURRENT AND VOLTAGE SATURATED FUNCTION SERIES

According to (3)-(5), a voltage saturated function series is described by (9) or its equivalent, a current saturated function series given by (10). Here, we can assume that the ratio of the resistors in Fig. **4** is Rf/Ri≥200. Then, the saturated shift h in Chapter 3 Fig. **4** is h_i=E_i. In this case, the comparing voltage and the saturated shift are almost the same. Based on this assumption, one can construct the more complex saturated function series given in (3.49) by using the basic cell in Fig. **4** with different comparing voltages E (E=0 and E≠0).

$$SFS(v) = F_S(v) + \sum_{i=1}^{N} F_S(v - Ei) + \sum_{i=1}^{N} F_S(v + Ei)$$

$$(9)$$

$$SFS(i) = \frac{SFS(v)}{R}$$

$$(10)$$

Therefore, in this book is proposed the general connection as shown in Fig. **5** to synthesize the current saturated function series in (10), and a modification of Fig. **5** to synthesize current and voltage saturated function series in (9) and (10) in the same circuit block as shown in Fig. **6**. Where BC in Figs. **5** and **6** is the basic cell notation, as shown in Fig. **4**, and *Cn* is the comparing voltage. The number of basic cells (BC) is determined by **BC=(scrolls-1)**. Furthermore, the structures in Fig. **5** and **6** are used to synthesize the saturated functions in Figs. **19**(b), **20**(b), **21**(b), **22**(b), **23**(b), **24**(b) for generating 1D-multi-scrolls attractors, 2D-multi-scrolls attractors and 3D-multi-scrolls attractors, respectively.

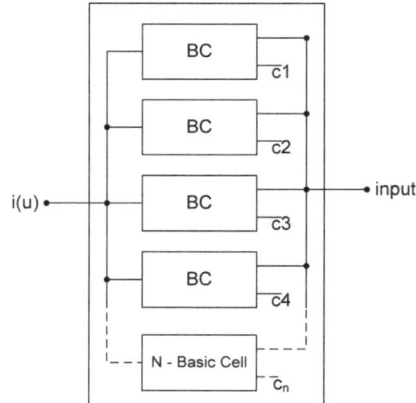

Figure 5: Structure to synthesize current saturated function series.

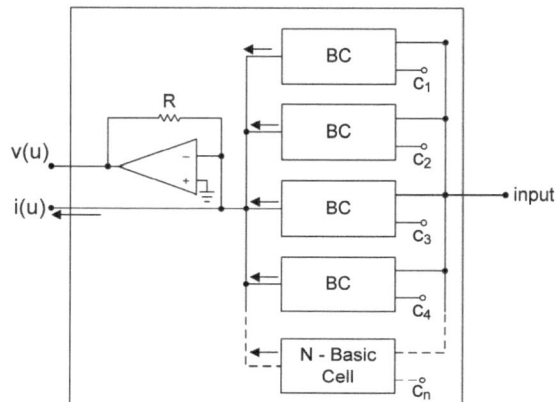

Figure 6: Structure to synthesize voltage and current saturated function series.

Saturated Functions for 1D-multi-scroll Chaotic Attractors

If Vsat=±2.5V (typical value for the commercially available opamp TLC2262 with Vdd=±2.6V [20]), Rix=10KΩ, Rc=100KΩ, Ri=1KΩ, Rf=1MΩ and E1=250mV, in (8), one gets k=250mV, Isat=25µA, α=2.5mV, s=100 and h=250mV. Furthermore, the synthesis result for the saturated function labeled SFx in Chapter 3 Fig. **19**(b) which generates a 1D-3-scroll chaotic attractor is shown in Fig. **7**(a). Similarly, the saturated function SFx in Chapter 3 Fig. **20**(b) that generates a 1D-6-scroll chaotic attractor is synthesized by setting E1=h1=500mV and E2=h2=1V in Fig. **4** as shown in Fig. **7**(b).

Saturated Functions for 2D-multi-scroll Chaotic Attractors

It should be pointed out that the synthesis process for the saturated function labeled by SFy in Fig **8** in both cases is realized by Fig. **6** to avoid the addition of another circuit block because the state variables system in (3.18) depends on two saturated functions SFy to generate scrolls on a 2D-mesh. In this manner, by selecting Vsat=±2.5V, Rix=R=10KΩ, Rc=100KΩ, Ri=1KΩ, Rf=1MΩ and E1=250mV in (8), one gets k=250mV, Isat=25µA, α=2.5mV, s=100 and h=250mV, which was used for the saturated functions SFx and SFy in Fig **21**(b) to generate 2D-3-scrolls. Besides, by changing E1=h=500mV, one gets the SFx and SFy in Chapter 3 Fig. **22**(b) to generate 2D-4-scrolls. The synthesis results are shown in Fig. **8**, respectively.

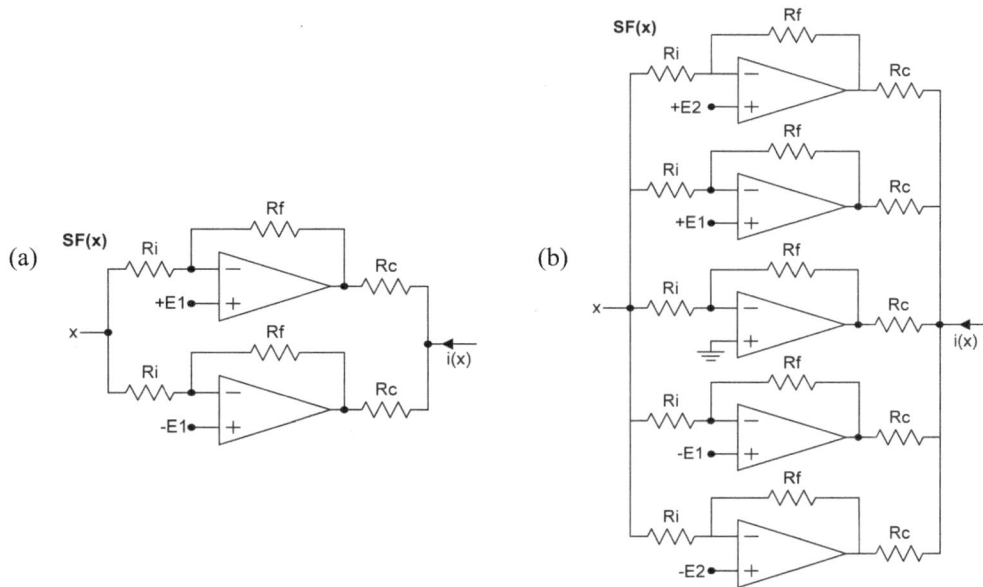

Figure 7: Opamp based-synthesis for: (a) saturated function in Chapter 3 Fig. **19**(b) with three plateaus and (b) saturated function in Chapter 3 Fig. **20**(b), with six plateaus.

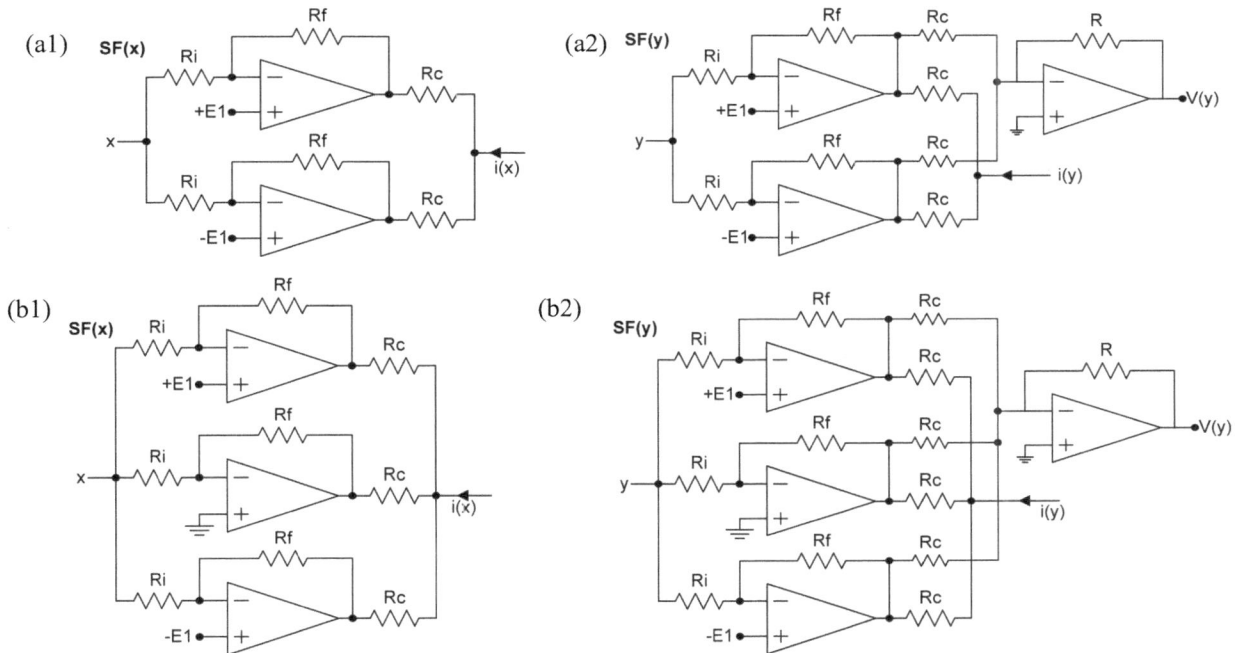

Figure 8: Opamp based-synthesis for: (a) saturated functions SFx and SFy in Chapter 3 Fig. **21**(b) with three plateaus and (b) saturated functions SFx and SFy in Chapter 3 Fig. **22**(b) with four plateaus.

Saturated Functions for 3D-multi-scroll Chaotic Attractors

Similarly, one can synthesize the saturated functions SFx, SFy and SFz in Chapter 3 Fig. **23**(b) to generate a 3D-3-scrolls attractor. By selecting Vsat=±10V (typical value for the commercially available opamp TL081[107]), Rix=R=10KΩ, Rc=100KΩ, Ri=1KΩ, Rf=1MΩ and E1=1V in (8) to synthesize SFx in Chapter 3 Fig. **23**(b); one gets k=1V, Isat=100μA, α=10mV, s=100 and h1=1V. Besides, the saturated functions SFy and SFz in Chapter 3 Fig. **23**(b) can be synthesized by changing: Vsat=±5V and E1=0.5V in (8) to get k=0.5V, Isat=50μA, α=5mV, s=100 and h1=0.5V. As mentioned above, SFx is synthesized by Fig. **5** while SFy and SFz are synthesized by Fig. **6** as shown in Fig. **9**.

Indeed, to synthesize the saturated functions SFx, SFy and SFz in Chapter 3 Fig. **24**(b) which generate a 3D-4-scrolls attractor, one only changes E1=2V to synthesize SFx and E1=1V to synthesize SFy and SFz. The resulting circuit is shown in Fig. **10**.

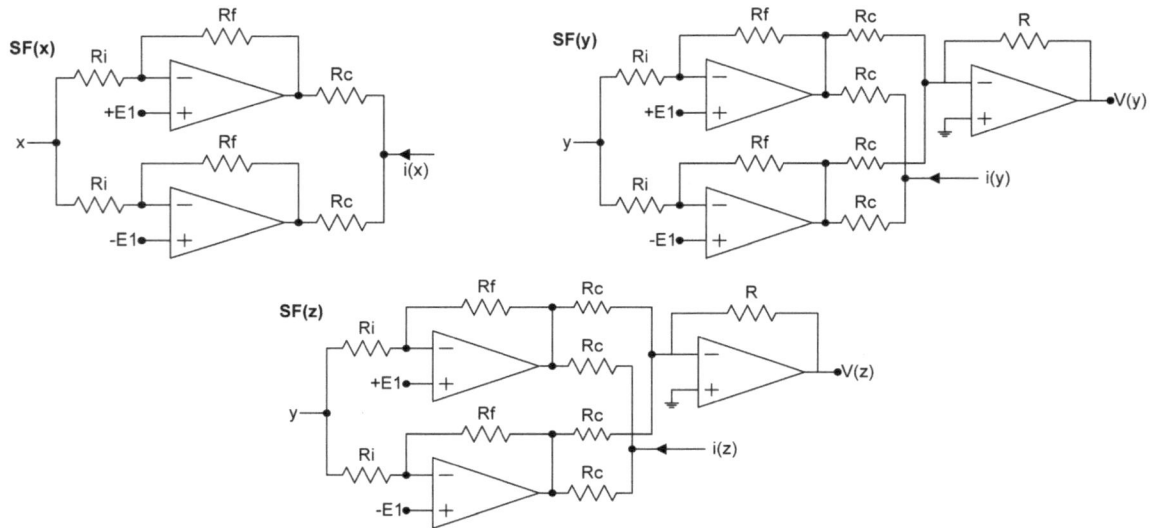

Figure 9: Opamp-based synthesis of SF(x), SF(y) and SF(z) in Chapter 3 Fig. **23**(b) with three plateaus.

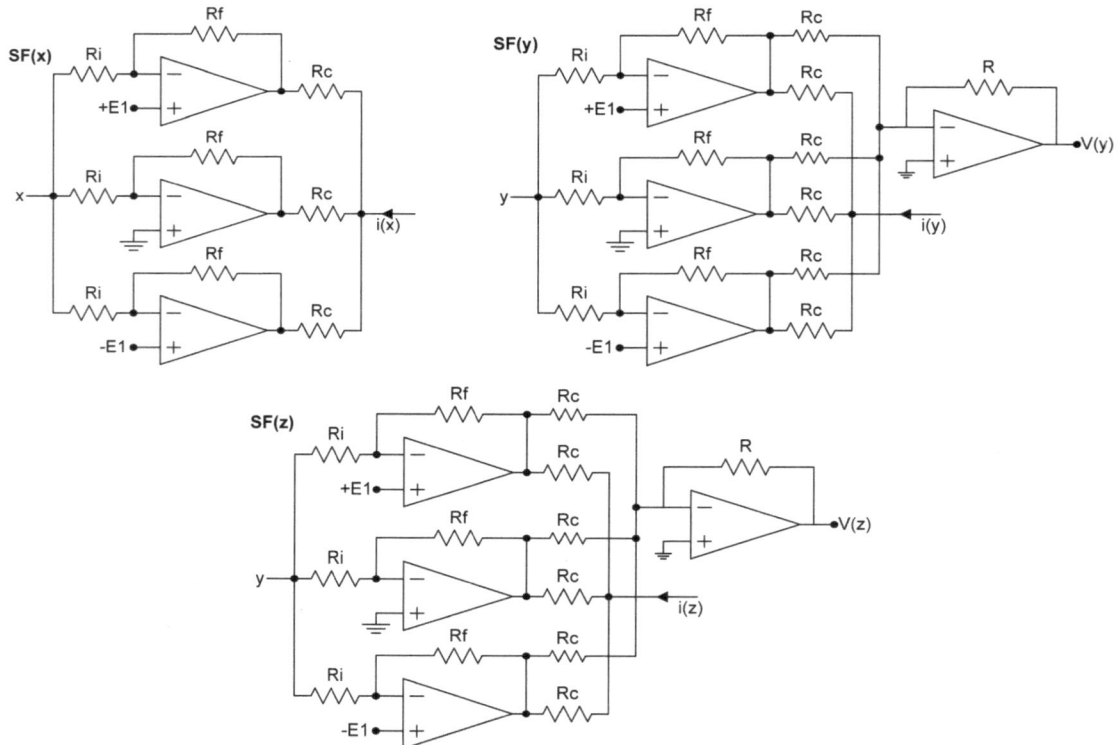

Figure 10: Opamp-based synthesis of SF(x), SF(y) and SF(z) in Chapter 3 Fig. **24**(b) with four plateaus.

OPERATIONAL AMPLIFIER VERILOG-A MODEL

A behavioural model describes the behaviour of a particular architecture by a set of mathematical relations between the input and output signals and the states of the system [90]. To further simplify the SPICE simulation into a top-down synthesis flow, a high-level opamp model is used to take into account ideal and non-ideal effects for the opamps in Fig. **7** to Fig. **10**. Therefore, analog signals can be handled more easily in a behavioural model than in circuit description [89]. The actual language used to describe the models depends on the concrete system that has to be designed. In this book, it is used the Verilog-A language [91].

A variety of modelling levels can be used to describe the opamps. These range from a simple functional model with a gain equation to complex models [90]. For instance, in this book it is used the model listed in Table **1** to study different effects on the opamps such as bandwidth BW, differential gain A_0, slew-rate SR and saturation effects. This model can be either refined or simplified according to the design requirements by applying the four operations defined in Chapter 2, subsection 2.4.2.

Table 1: Verilog-A model.

```
module opa(inp, inn, out, vdd, vss);
input inp, inn, vdd, vss;
inout out;
electrical inp, inn, out, vdd, vss, vin, vo;
parameter real Rm = 2.0k, Cm = 20p, G = 250k;
real v;
```

```
analog begin
V(vin) <+ laplace_nd(G*V(inp,inn), {1.0},{1.0, 1.0e-6});
I(vin,vo) <+ V(vin,vo)/Rm;
I(vo) <+ ddt(Cm*V(vo));
v = V(vo);
if ( v > V(vdd) ) begin V(out) <+ V(vdd); end
else if ( v < V(vss) ) begin V(out) <+ V(vss); end
else begin V(out) <+ v; end
end
endmodule
```

SPICE SIMULATION RESULTS

This section presents the SPICE simulation results for the synthesis process of the saturated functions in Fig. **7** to **10** using the verilog-A model introduced in the previous section. Note that, in Fig. **11** to **15** the value for the saturated plateaus is given either in current or in voltage.

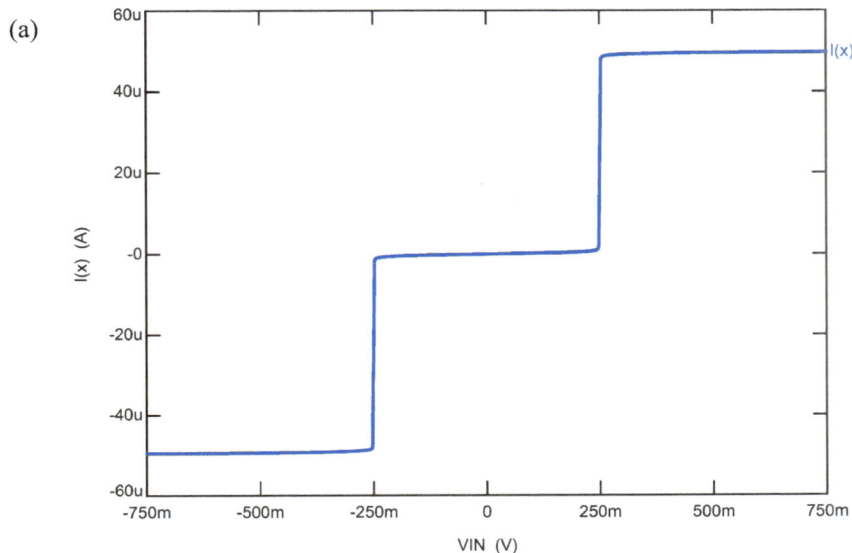

(a)

(b)

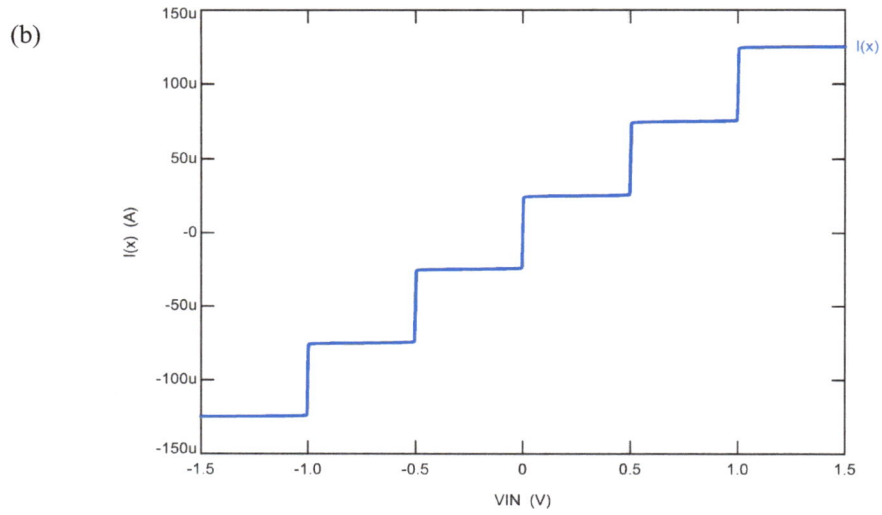

Figure 11: SPICE results for: (a) current saturated function in Fig. **7**(a) and (b) current saturated function in Fig. **7**(b).

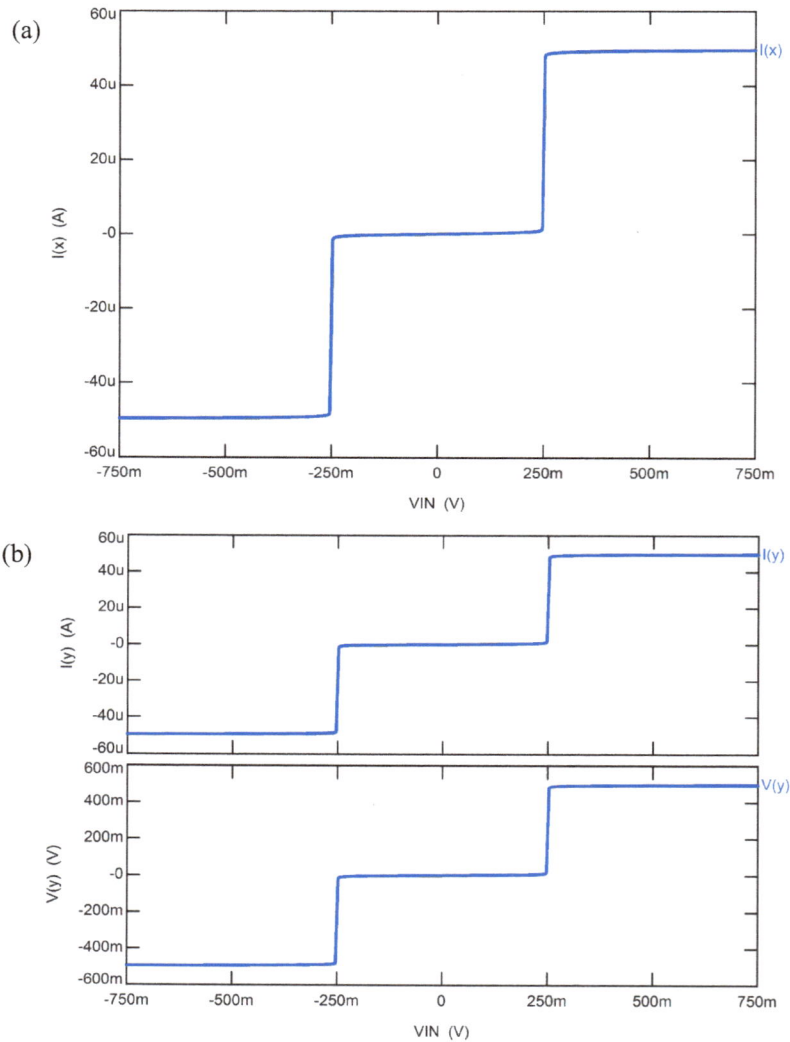

(a)

(b)

Figure 12: SPICE results for: (a) current saturated function SFx and (b) voltage and current saturated function SFy in Fig. **8**(a), respectively.

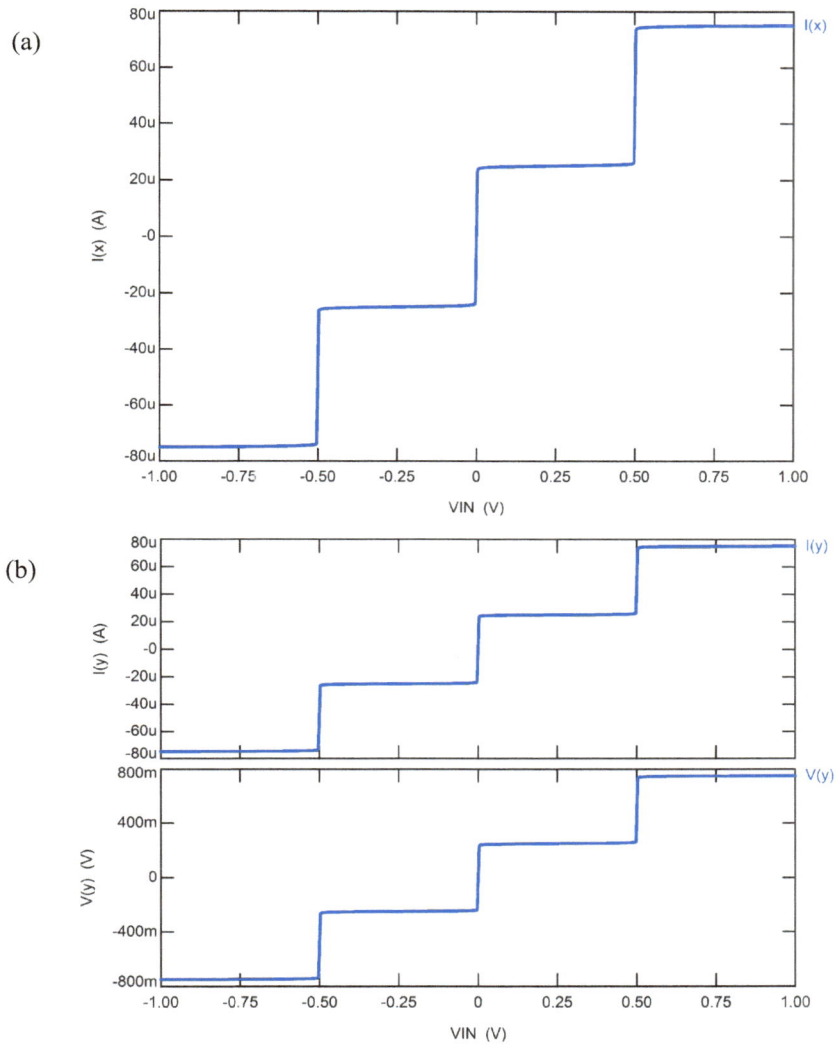

Figure 13: SPICE results for: (a) current saturated function SFx and (b) voltage and current saturated function SFy in Fig. **8**(b), respectively.

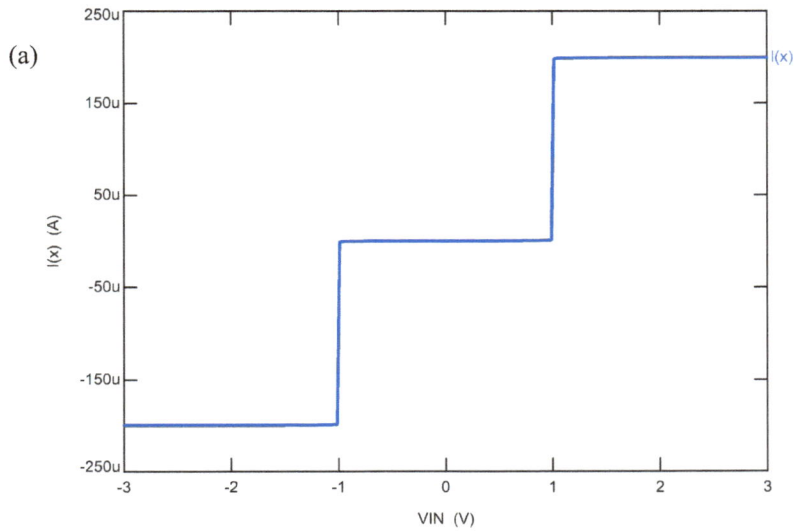

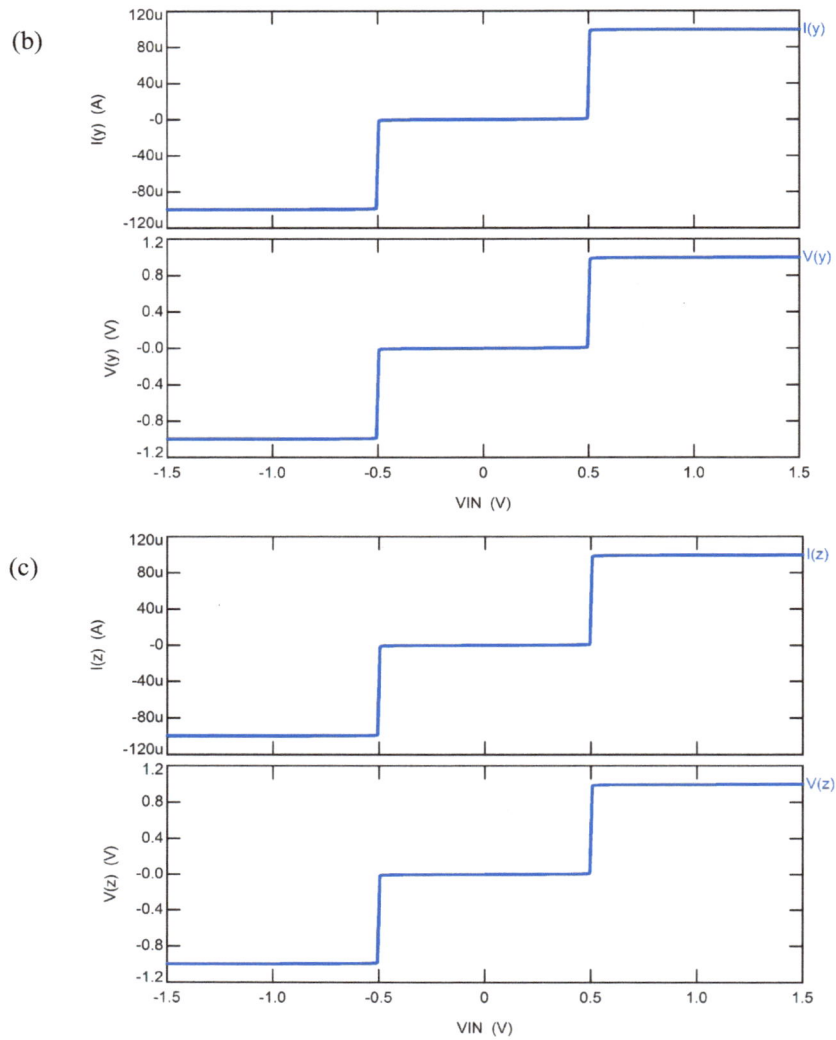

Figure 14: SPICE results for: (a) current saturated function SFx, (b) voltage and current saturated function SFy and (c) voltage and current saturated function SFz in Fig. **9**, respectively.

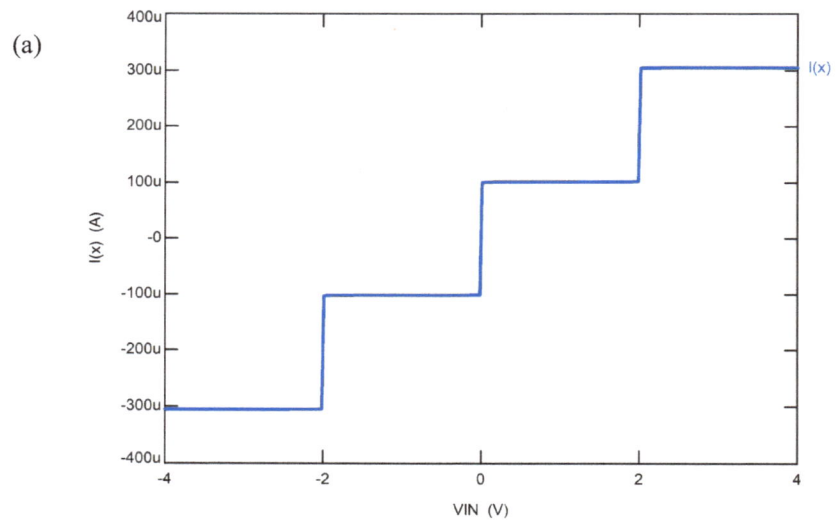

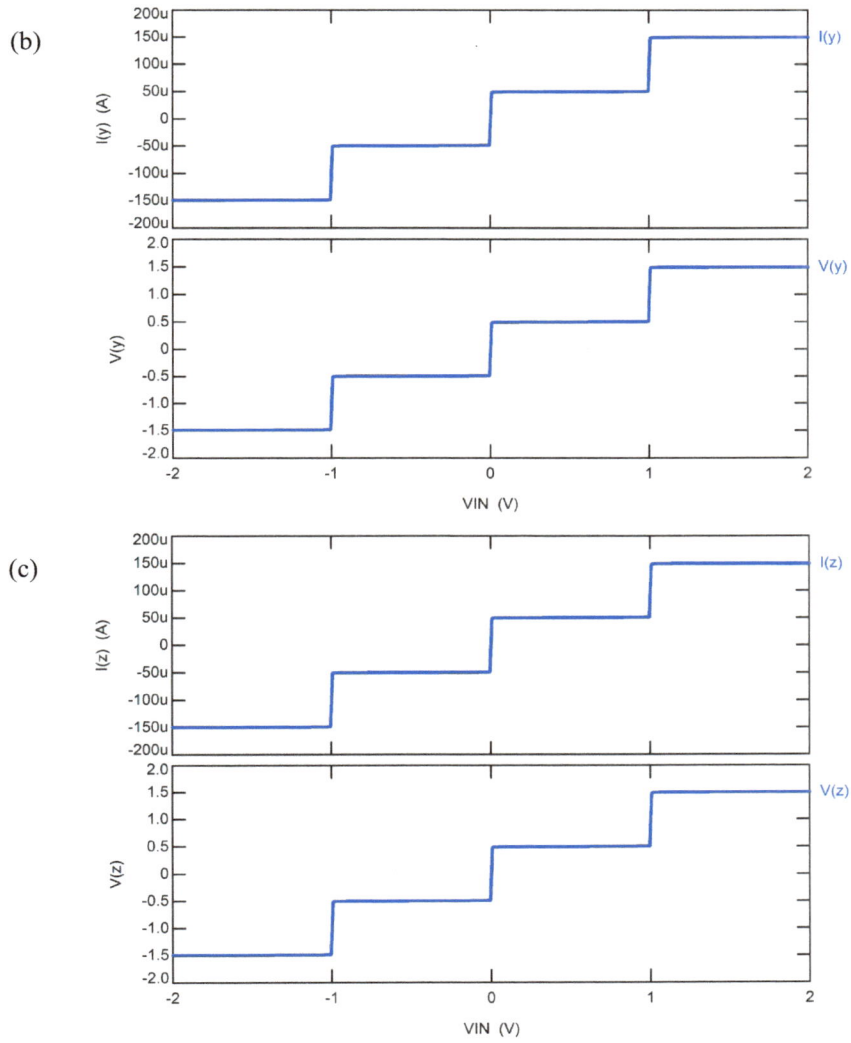

Figure 15: SPICE results for: (a) current saturated function SFx, (b) voltage and current saturated function SFy and (c) voltage and current saturated function SFz in Fig. **10**, respectively.

CHAPTER 5

Synthesis Methodology for Multi-Scrolls Chaotic Attractors

Abstract: The behavioural modelling strategy based on state variables and PWL approximations allows us to represent several chaotic systems at the electronic system level (ESL) of abstraction. The design space exploration derived from numerical simulation approaches in the time domain can be adopted to estimate the performance values of the chaotic systems. That is only one part of a design strategy. Indeed, a question remains: which architecture and parameters values must be selected during the refinement operation of the behavioural models? Therefore, the synthesis methodology presented in this chapter completes the opamps based-synthesis of saturated functions introduced in Chap. 4. First, the steps of a new synthesis methodology proposed in this book for the design of chaotic systems based on behavioural modelling are summarized. The new methodology of circuit synthesis is performed by three hierarchical levels. Finally, numerical results are confirmed by H-SPICE simulations to show the usefulness of the proposed synthesis approach.

Keywords: Chaos, electronic design automation, dynamical systems, chaos generators, modelling and simulation.

SYNTHESIS METHODOLOGY

This section introduces the guidelines to synthesize multi-scrolls chaotic systems by means of high-level descriptions. The aim of this investigation is to synthesize multi-scrolls chaotic systems based on saturated functions with multi-segments. The new methodology of circuit synthesis is performed by three hierarchical levels in a top-down fashion as shown in Fig. **1**. Therefore, first the objectives and basic properties of the synthesis strategy developed in this work are defined.

The synthesis methodology operates mainly on the behavioural description level, but at different levels of abstraction. The main goal of the synthesis methodology is to convert a functional description of a complex analog system (chaotic system) into a circuit description at a lower abstraction level of a specific topology. More specifically, seven objectives form the structure of the synthesis approach.

1. The behavioural modelling strategy adopts a top-down design methodology starting from a functional description level to cope with systems with a higher complexity (chaotic systems).

2. Circuit representations for saturated functions and chaotic systems can be created as the result of the behavioural modelling process.

3. The algorithm performs an estimation of both saturated functions topology and parameter values concurrently.

4. The process deals with different types of performance characteristics of the chaotic oscillators generated by the user.

5. Excursion levels and frequency scaling are taken into account and used to explore the design process towards synthesis of the saturated functions using electronic devices.

6. The synthesis methodology defines several tasks which can be quickly executed by using automatic numerical approaches at higher levels of abstraction and hardware description language (HDL) models at lower levels of abstraction

7. The synthesis methodology of multi-scrolls chaotic systems leads to develop a simple high-level solution into an analog circuit describing the entire chaotic system at a low abstraction level (opamp-level).

These objectives are the foundations of the top-down hierarchical synthesis methodology presented in this book. The general flow is shown in Fig. **1** and it starts with high-level descriptions, which capture the behaviour of the multi-scrolls chaotic attractors and these include the number of scrolls, position of the scrolls on x-axis, y-axis and z-axis, voltage or current level of the chaotic signals and frequency of the attractor. The procedures at each hierarchical level are described below.

Jesus Manuel Muñoz Pacheco and Esteban Tlelo Cuautle

LEVEL 1

- *High-level behavioural modelling*: It is the highest level in the hierarchy of the synthesis methodology. The chaotic systems are modelled by applying state variables approach and PWL approximations as shown in Chapter 3. Therefore, the strong nonlinear behaviour of the chaotic systems is represented by high-level descriptions, as a result; one can explore the design space for multi-scrolls chaotic systems based on saturated functions. Consequently, designers can verify its designs before the physical implementation with electronic circuits to avoid time-waste when using complex SPICE models.

- *Excursion levels and frequency scaling*: The procedure of excursion levels (ELs) scaling is executed at this level because original nonlinear functions have ELs outside the range that real opamps can handle. Thereby, the ELs of the chaotic signals are scaled to control the breaking points and slopes of the saturated functions within practical values as shown in Chapter 3. Additionally, the frequency scaling of multi-scrolls chaotic attractors is performed. In this manner, the attractor frequency is only limited by the bandwidth of real opamps.

- *Numerical simulation*: It is used to estimate the values of the circuit elements. To speed-up time simulation, the chaotic systems are numerically simulated at the ESL by applying state variables and PWL approximation and by executing automatic control and determination of step-size for multi-step algorithms. This step completes the previous ones and together forms the first level of hierarchy.

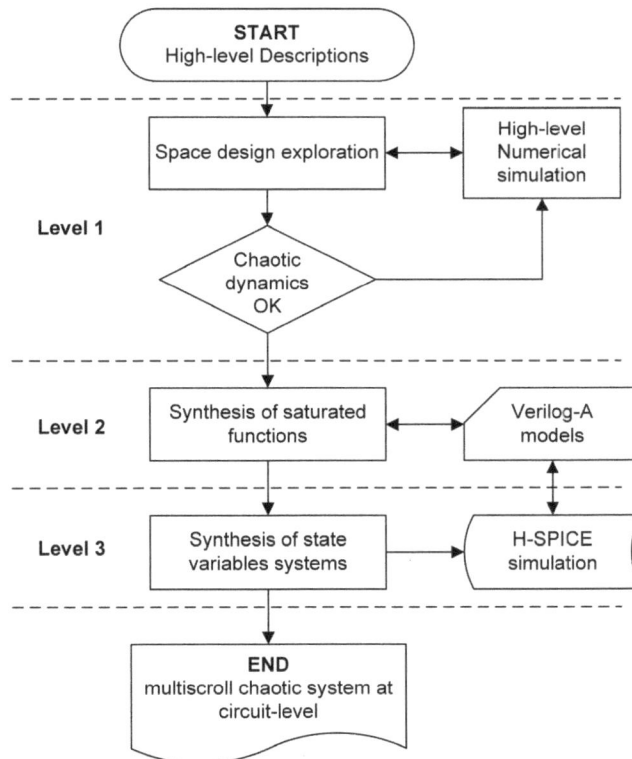

Figure 1: High-level analog synthesis environment.

LEVEL 2

- *Synthesis of PWL functions*: The PWL functions are synthesized using a basic cell interconnected in order to design voltage and current saturated functions series as shown in chapter 4. Therefore, the high-level parameters of the saturated functions are carried out to the next level of hierarchy by means of a refinement operation.

- *Non-ideal effects*: Verilog-A models are generated to include limitations of real opamps, like gain (Ao), bandwidth, slew rate, saturation. Therefore, the performance of the saturated functions is evaluated by SPICE and Verilog-A simulations.

LEVEL 3

- At this level, the state variables systems are also synthesized with opamps, which are selected from an analog component library according to restrictions imposed by the first level of hierarchy. In this manner, a chaotic system is designed using opamps to generate multi-scrolls chaotic attractors with multiple orientations (1D, 2D or 3D). High-level numerical simulation results are confirmed by H-SPICE simulations to show the usefulness of the proposed synthesis approach.

Furthermore, behavioural modelling is herein exploited to give a solution on the synthesis of chaotic systems, because it offers one possible way to abstract the features of interest in a circuit block or a system.

SYNTHESIS OF MULTI-SCROLLS CHAOTIC ATTRACTORS

This section shows how the synthesis methodology introduced in the previous section can synthesize multi-scrolls chaotic attractors by using behavioural modelling.

1D-multi-scrolls-chaotic Attractors

The state variables system in Chapter 3 Equation (17) which generates 1D-chaotic attractors, can be also synthesized with opamps as shown herein. Chapter 3 Equation (17) has the block diagram representation given in Fig. **2** that is realized with three integrators. Each block can be synthesized with opamps, as shown in Fig. **2**. The operator (+) is realized by adding currents in a node. Note that, a current saturated function series is used in this implementation since it is necessary to add current signals in the same node. The analog circuit for the system in Chapter 3 Equation (17) is shown in Fig. **3**. By applying Kirchhoff's laws in Fig. **3** one obtains (1), where the current saturated function SFx=i(x)Rix. The synthesis parameters in (1) are determined by (2). If Rix=10KΩ, then C$\approx$143µF, Rp=7KΩ, Rx=Ry=Rz=10KΩ, Rn=10KΩ and Rm=10KΩ, which describes the same system given in Chapter 3 Equation (17). At this level, it is also possible to apply the frequency scaling procedure and it is done on capacitor C in (1) by evaluating Chapter 3 Equation (52). The frequency scaling is limited by the finite bandwidth of the opamps in Fig. **3**.

$$\frac{dx}{dt}=\frac{y}{RpC} \quad \frac{dy}{dt}=\frac{z}{RpC} \quad \frac{dz}{dt}=-\frac{x}{RxC}-\frac{y}{RyC}-\frac{z}{RzC}+\frac{[i(x)Rix]}{RixC} \tag{1}$$

$$C=1/0.7Rix \quad Rx=Ry=Rz=1/0.7C \quad Rp=1/C \tag{2}$$

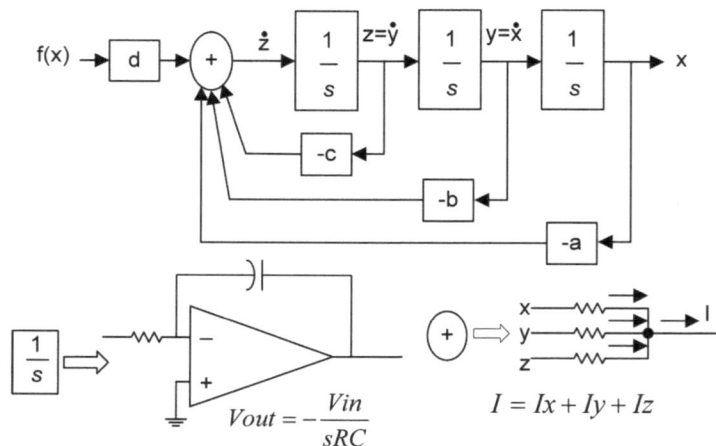

Figure 2: Block diagram description of (3.17).

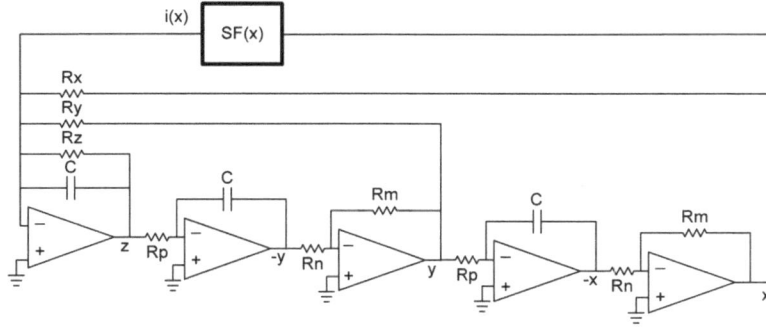

Figure 3: Opamp-based implementation of (3.17).

2D-multi-scrolls-chaotic Attractors

Similarly, the state variables system in (3.18) which generates 2D-chaotic attractors is synthesized with opamps where it has the block diagram representation given in Fig. **4** that is also realized with three integrators. The main difference with the previous case is that two saturated functions are used in this implementation, the first for generating the current saturated function series and the second for generating voltage and current saturated function series. Note that, a voltage saturated function series is necessary to realize the subtraction operation in the state variable x. The analog circuit for the system in (3.18) is shown in Fig. **5**. By applying Kirchhoff´s laws in Fig. **5** one obtains (3), where the current saturated function SFx=i(x)Rix, the current saturated function SFy=i(y)Rix and the voltage saturated function SFy=v(y). The synthesis parameters in (3) are also determined by (2). If Rix=10KΩ, then C≈143μF, Rp=7KΩ, Rx=Ry=Rz=10KΩ, Rn=10KΩ and Rm=10KΩ, which describes the same system given in (3.18). At this level, it is also possible to apply the frequency scaling procedure and it is done on capacitor C in (3) by evaluating (3.52). The frequency scaling is only limited by the finite bandwidth of the opamps in Fig. **5**.

$$\frac{dx}{dt}=\frac{y}{RpC}-\frac{v(y)}{RpC} \qquad \frac{dy}{dt}=\frac{z}{RpC} \qquad \frac{dz}{dt}=-\frac{x}{RxC}-\frac{y}{RyC}-\frac{z}{RzC}+\frac{\left[i(x)Rix\right]}{RixC}+\frac{\left[i(y)Rix\right]}{RixC}$$

(3)

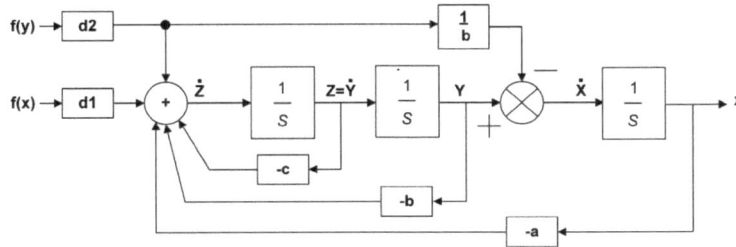

Figure 4: Block diagram description of (3.18).

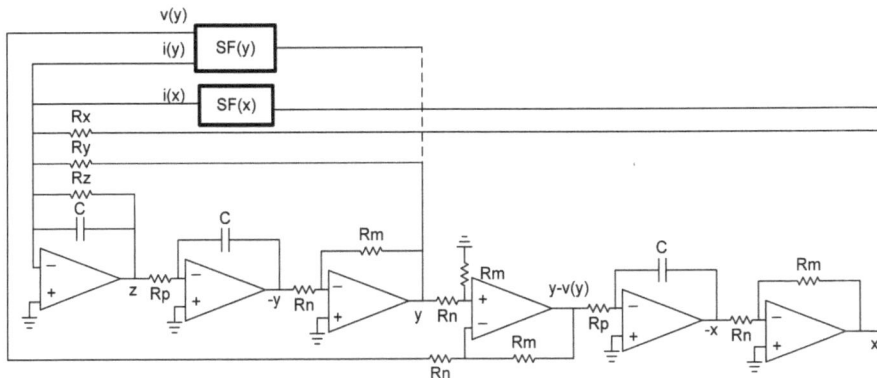

Figure 5: Opamp-based implementation of (3.18).

3D-multi-scrolls-chaotic Attractors

Finally, the state variables system in (3.24) which generates 3D-chaotic attractors, is synthesized with opamps where it has the block diagram representation given in Fig. **6** that is again realized with three integrators. The main difference with the previous case is that three saturated functions are used in this implementation, the first for generating a current saturated function series and the second and third for generating voltage and current saturated function series. Note that, two voltage saturated function series are necessary to realize the subtraction operation in the state variables x and y. The analog circuit for system in (3.24) is shown in Fig. **7**.

By applying Kirchhoff's laws in Fig. **7** one obtains (4), where the current saturated function SFx=i(x)Rix, the current saturated function SFy=i(y)Rix, current saturated function SFz= i(z)Rix, the voltage saturated function SFy=v(y) and the voltage saturated function SFz=v(z). The synthesis parameters in (3) are again determined by (2). If Rix=10KΩ, then C≈143μF, Rp=7KΩ, Rx=Ry=Rz=10KΩ, Rn=10KΩ and Rm=10KΩ, which describes the same system given in (3.24). The frequency scaling is done on capacitor C in (4) by evaluating (3.52). As for the previous cases, the frequency scaling is only limited by the finite bandwidth of the opamps in Fig. **7**.

$$\frac{dx}{dt}=\frac{y}{RpC}-\frac{v(y)}{RpC} \qquad \frac{dy}{dt}=\frac{z}{RpC}-\frac{v(z)}{RpC} \qquad \frac{dz}{dt}=-\frac{x}{RxC}-\frac{y}{RyC}-\frac{z}{RzC}+\frac{[i(x)Rix]}{RixC}+\frac{[i(y)Rix]}{RixC}+\frac{[i(z)Rix]}{RixC} \qquad (4)$$

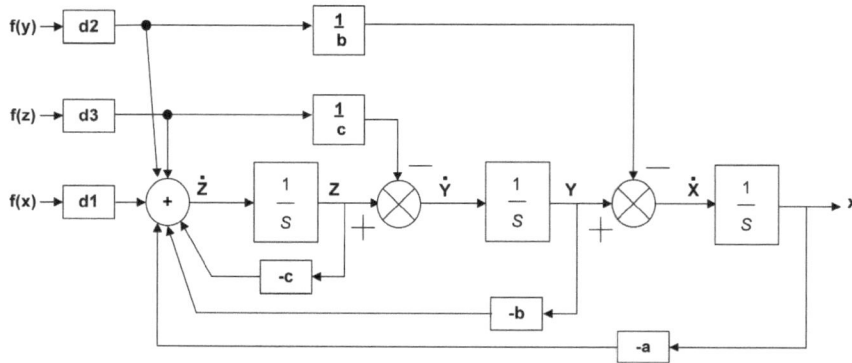

Figure 6: Block diagram description of (3.24).

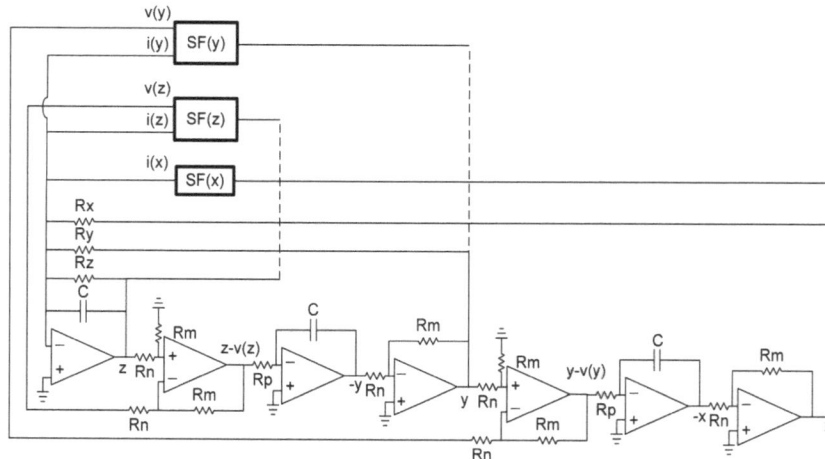

Figure 7: Opamp-based implementation of (3.24).

SPICE SIMULATION RESULTS

To demonstrate the usefulness of the synthesis methodology in analog design, it is used herein to synthesize the multi-scrolls systems previously simulated in Chapter 3 at the ESL by applying behavioural modelling. For all cases, this section shows the synthesized chaotic attractors with an even and odd number of scrolls. Furthermore, the synthesis results for 3 and 6 scrolls chaotic attractors with 1D direction are shown in Fig. **8**, by combining the saturated function series in Chapter 4 Fig. **7**(a, b) (previously synthesized in Chapter 4) with Fig. **3** to generate 3 and 6-scrolls, respectively.

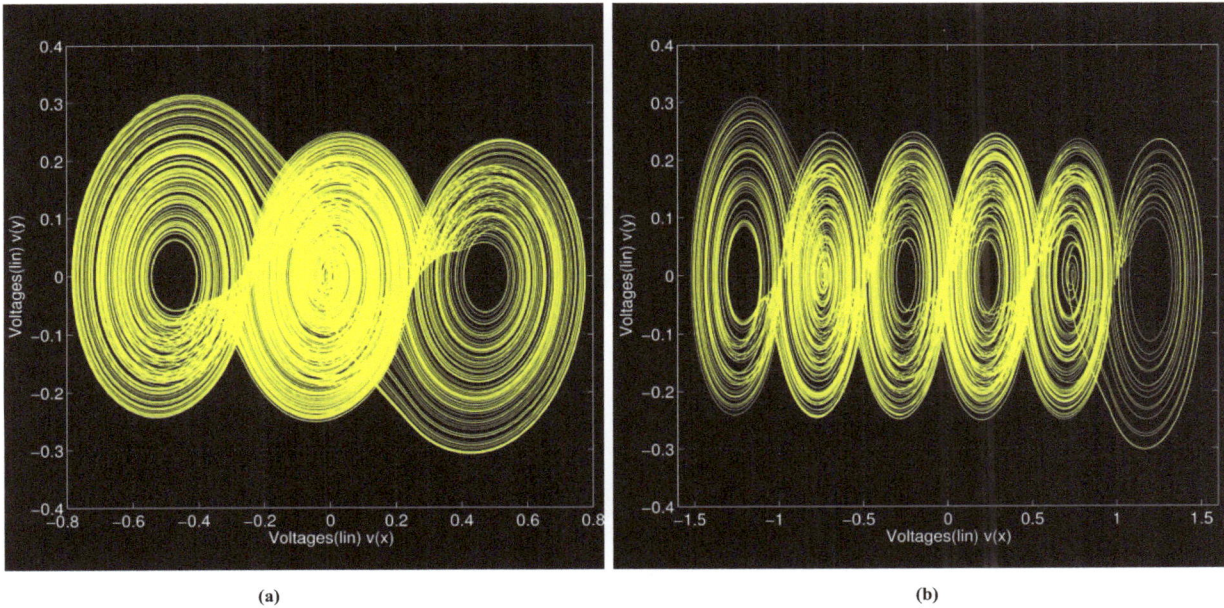

(a)   (b)

Figure 8: H-Spice results for 1D chaotic attractors.

Additionally , the synthesis results for 3 and 4 scrolls chaotic attractors with 2D direction are shown in Fig. **9**, by combining the saturated function series in chapter 4 Fig. **8**(a, b) (previously synthesized in Chapter 4) with Fig. **5** to generate 3 and 4-scrolls, respectively.

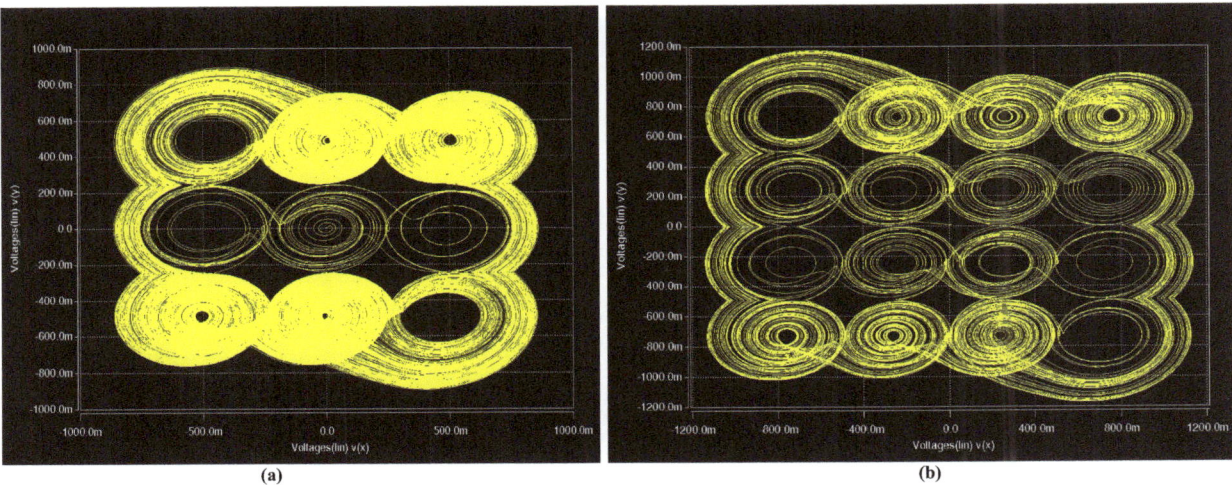

(a)   (b)

Figure 9: H-Spice results for 2D chaotic attractors.

Finally, the synthesis results for 3 and 4 scrolls chaotic attractors with 3D direction are shown in Figs. **10** and **11**, by combining the saturated function series in Chapter 4 Figs. **9** and **10** (previously synthesized in Chapter 4) with Fig. **7**, to generate 3 and 4-scrolls, respectively. By selecting a frequency scaling factor one gets Csf, the new value for C in (4). For C=143µF, the original value, the frequency spectrum of Fig. **11** is shown in Fig. **12**(a) where $fout$=0.154 Hz. For the scaling factors of 65000, 325000 and 650000, C is updated to 2.2nF, 440pF and 220pF, respectively, as result $fout$=10kHz, $fout$=50kHz and $fout$=100kHz, as shown in Fig. **12**(b) to (d), respectively.

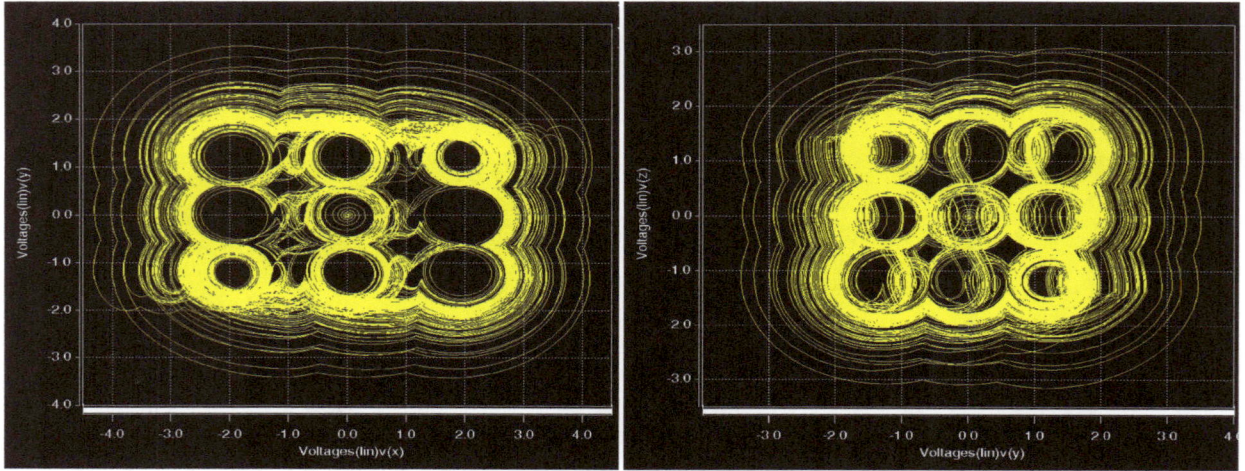

(a)   (b)

Figure 10: H-Spice results for 3D chaotic attractors with odd-scrolls.

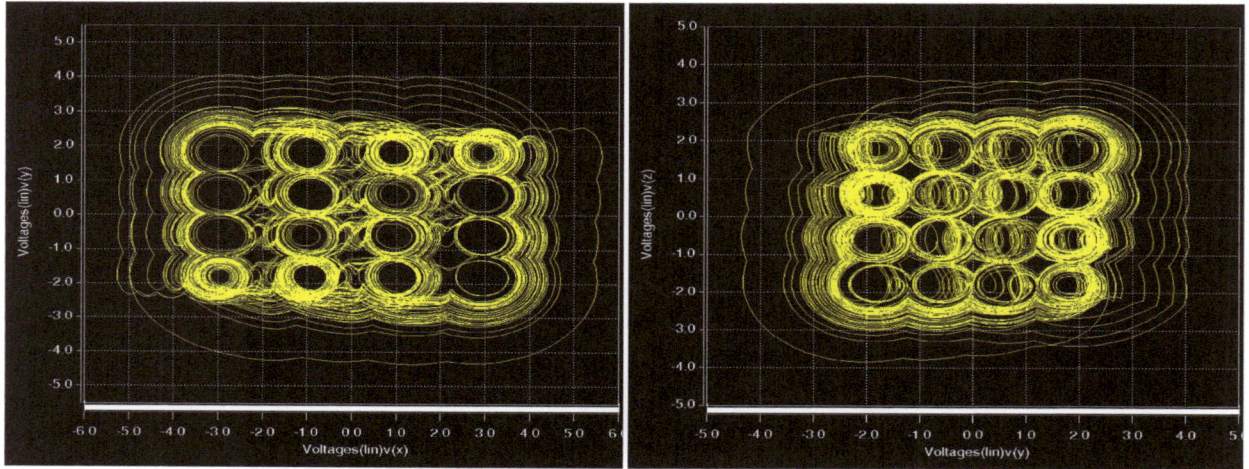

(a)   (b)

Figure 11: H-Spice results for 3D chaotic attractors with even-scrolls.

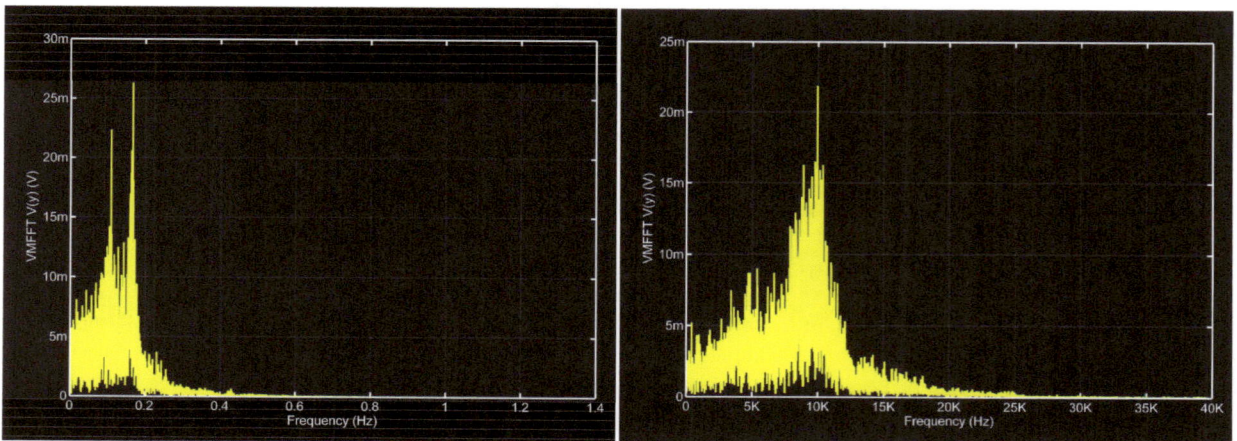

(a)   (b)

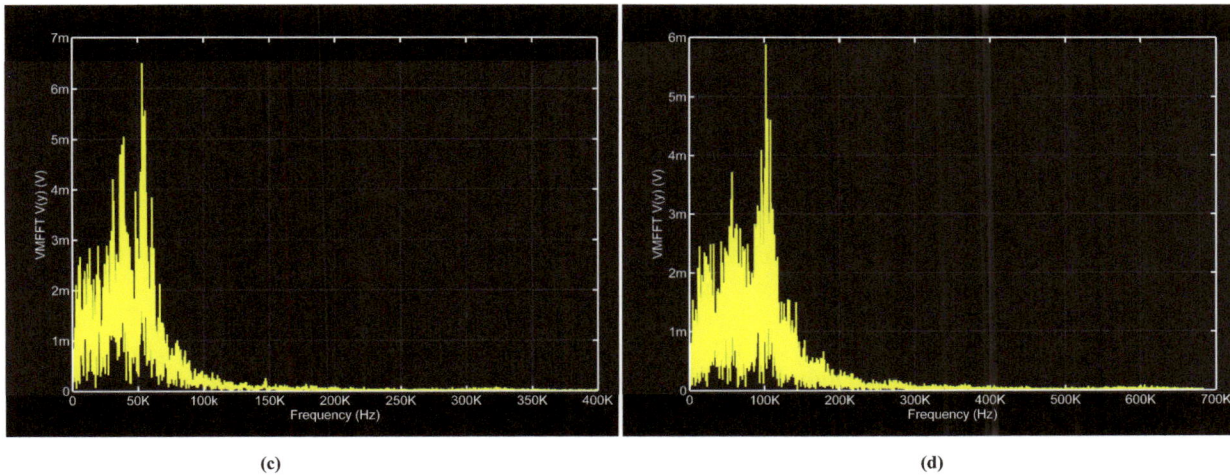

(c)   (d)

Figure 12: (a) Frequency spectrum of Fig. **11** and scaling by (b) 65000 (f=10KHz), (c) 325000 (f=50KHz) and (d) 650000 (f=100KHz).

AN EDA TOOL PROTOTYPE

The top-down hierarchical design flow for 1D, 2D and 3D-multi-scrolls chaotic systems based on saturated functions (SFs) is supported on this tool by beginning with Electronic System Level (ESL) simulations, and ending with the synthesis of each individual block using operational amplifiers (opamps) [134,135]. The principal advantage of this electronic design automation (EDA)-tool compared to others [32] is that it offers the possibility to analyze and design chaotic systems from behavioural modelling (ESL level) to electronic devices (physical level). It does not exist previously works that cover the entire analog design flow for chaotic systems using behavioural modelling [7,20].

The design of multi-scrolls chaotic systems starts with a high-level description of its desired behaviour and the EDA-tool creates a circuit that satisfies the specified design goals. Besides, it determines the sizing (numerical values) of all of the components for the chaotic oscillator and saturated functions. To facilitate that, the tool can provide to the user with a graphical-user-interface (GUI) that allows him to design chaotic systems with a minimal input (number of scrolls, position of the scrolls on x-axis, y-axis and z-axis for 1D, 2D and 3D, voltage or current level of the chaotic signals and frequency of the attractor) [101,134,135]. Therefore, this section reviews the operation of the tool and describes how it can be used to design multi-scrolls chaotic attractors using opamps. The operation of the CAD-tool introduced in this book is executed by different procedures as shown in a simplified flowchart in Fig. **13**(a) and described below.

- A set of numerical routines containing the implementation of 3th-order Adams-Moulton algorithm (previously introduced in Chapter 3) designed for the numerical integration task. The advantages of these numerical algorithms are that both determination and control of step-size are automatically calculated according to the minimum absolute value of all the eigenvalues of the state matrix. Consequently, the time simulation is reduced by combining these algorithms with state variables and piecewise linear (PWL) approximations.

- A set of structural routines containing the functions designed to translate the behavioural descriptions of the chaotic systems into an analog circuit using opamps. This EDA-tool is further incorporating other kind of electronic devices, for instance: current-feedback operational amplifiers (CFOAs) [28,106] and current conveyors (CC) [31,101]. The translation is easily done by means of a direct mapping between the slopes and breakpoints of the saturated functions and the general connection of opamp-based basic cells as shown in Chapter 4 by using the opamp finite gain model. At this level, the EDA-tool generates a netlist and the users can execute a full SPICE simulation using opamp high-level Verilog-A or SPICE models taken from a library. Those models can be written by the users according to the real effects that they want to take into account or downloaded from the opamp vendor website. In this book, we use the Verilog-A model shown in Chapter 4 Table **1** and the SPICE macro-model for the commercially available opamps TL081 [107] and TLC2262 [108].

- A GUI. For instance, *Maple* provides a mechanism to generate GUIs by using the so-called *Maplet* [134,135]. Although its implementation requires a high level of expertise, it allows to end users to apply the proposed synthesis approach with a minimal knowledge and only five input parameters, thus facilitating its use and dissemination.

The EDA-tool for synthesizing multi-scroll chaos generators can be implemented in various languages. In [136] the tool was implemented in *Maple*™ by using a Pentium IV processor at 3.0 GHz, with 1.0 GB of RAM. Therefore, from Figs. **14** and **16** are shown some screen snapshots of the different windows of the application. The "designs manager" window shown in Fig. **13**(b) controls the main windows for the design of 1D, 2D and 3D chaotic attractors. The main windows of the tool, from where the other windows will show up when invoked are shown in Fig. **14**.

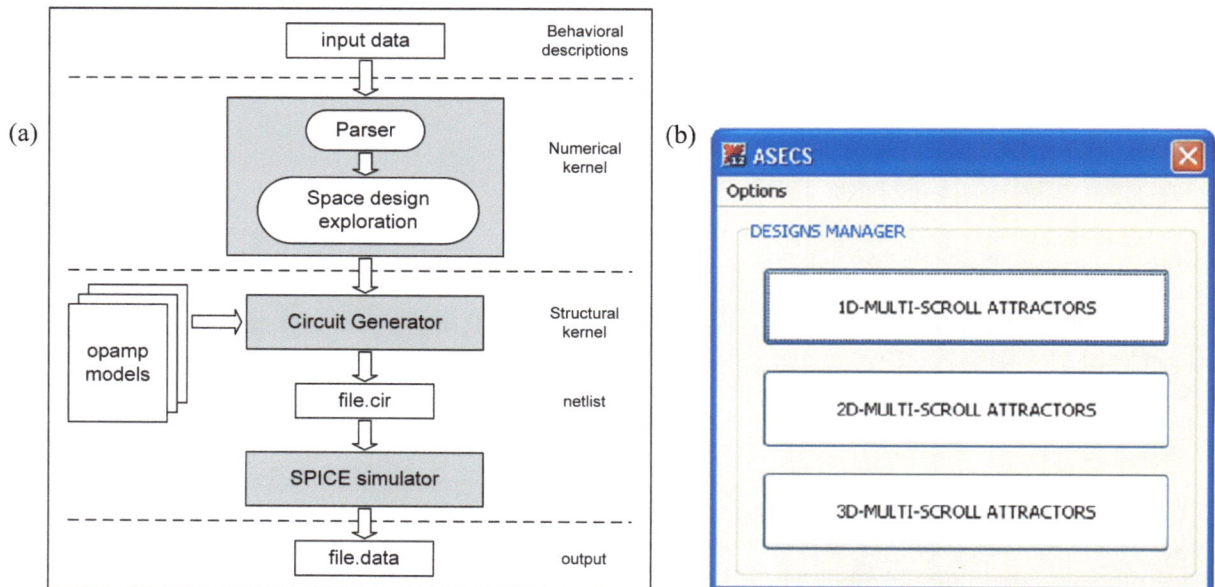

Figure 13: (a) EDA-tool workflow and (b) Screenshot of the Designs Manager window.

The workflow to design multi-scrolls chaotic systems is as follows: When the user enters the behavioural parameters (number of scrolls, position of the scrolls, voltage or current level of the chaotic signals and frequency of the attractor) [101, 134-136], the tool checks chaos conditions and computes the numerical solution for the chaotic system. At this stage, the three-dimensional chaotic dynamical systems are modelled by applying states variables and PWL approximations. From these windows, the user can control both synthesis options and high-level parameters for the chaotic systems.

The EDA tool includes a "NEW CASE" button, when it is activated, all memory records will be deleted. Any message about the synthesis process is shown in the "feedback pane" at the bottom in Fig. **14**. By pressing the "BEHAVIOURAL" button, windows in Fig. **15** are invoked and it shows the graphical representation of the attractor, the saturated functions and the evolution of state variables over the time. Consequently, the designer can relate the high-level parameters with the chaotic behaviour of the system. This process is iteratively repeated until the performance of the chaotic system agrees with the design requirements. It should be pointed out that the chaotic exploration at this level is realized by applying behavioural modelling originating that the time simulation is speeded up.

The "CIRCUIT" button opens the window in Fig. **16**(a) where different opamp models (including the frequency scaling factor leading to window in Fig. **16**(b)) can be chosen. When the opamp model and the frequency scaling factor have been introduced, the "GENERATE" button computes the connections between active and passive components and determines the sizing (numerical values) of all components for the chaotic oscillator and saturated functions. The final results are the graphical output shown in the window of Fig. **17** where an opamp-based analog circuit for the state variables system and the saturated functions is displayed. Besides, the CAD-tool generates a netlist to be used into a SPICE simulation of the multi-scrolls chaotic system.

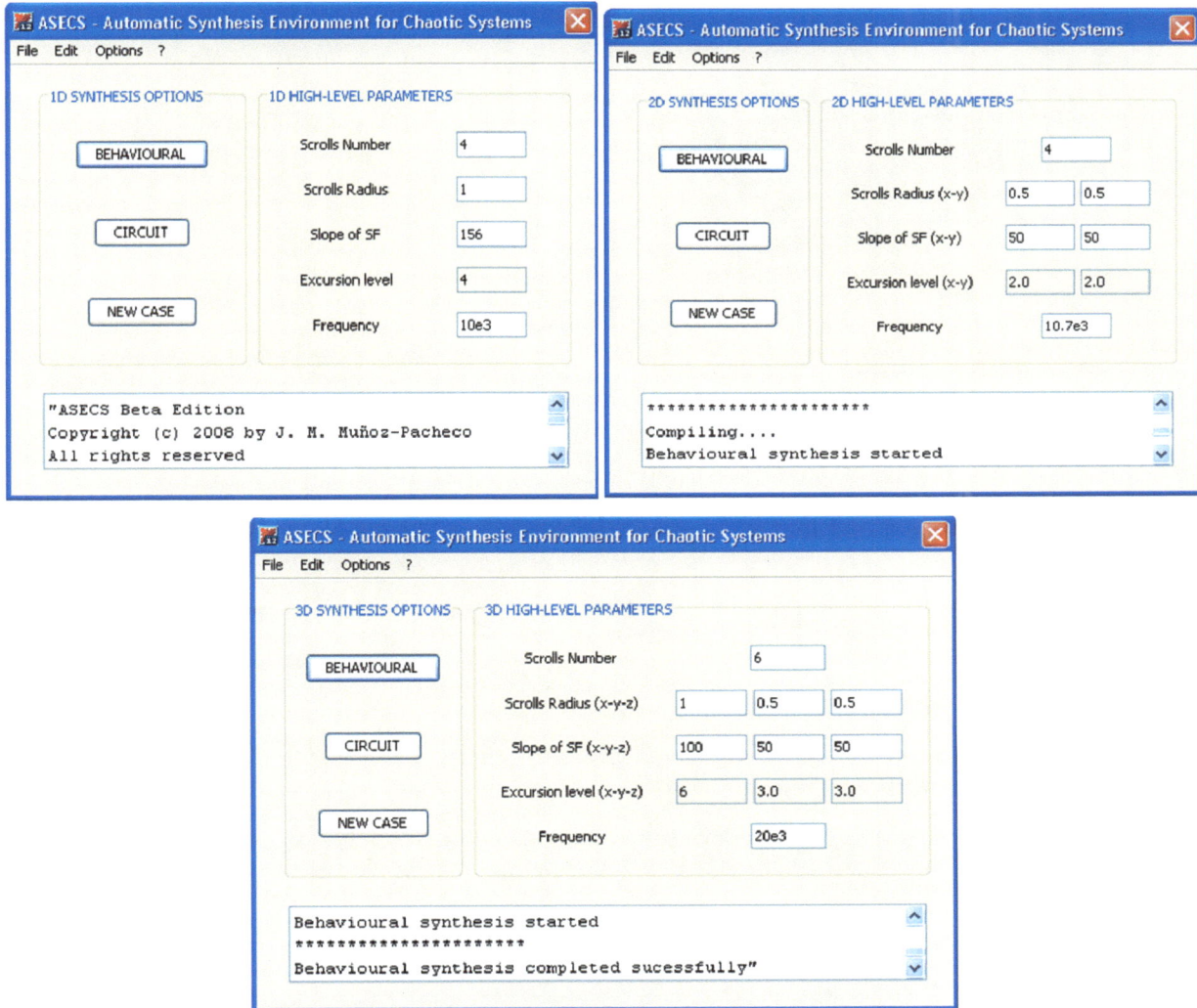

Figure 14: Screenshots of the windows for the EDA-tool developed in [136].

CASE STUDY

To demonstrate its utility in the area of analog circuit design, the EDA-tool was used to design a 1D-4-scrolls chaotic attractor. First, the user declares the high-level parameters as shown in Fig. **14**(a). In this example, they are considered "number of scrolls" = 4, "scroll radius" = 1V, "slope of SFx" = 156, "EL" = 4V and "frequency" = 10KHz. By pressing the "BEHAVIOURAL" button, the tool computes the chaotic system parameters in (3.17) and (3.49), obtaining k1=1, α1=6.4e-3, h1=2, p1=q1=1 for the saturated function. As a result, a 1D-4-scrolls chaotic attractor and its saturated function SFx along with the evolution of the system variables over the time are shown in Fig. **15**. The next step is to synthesize the state variables system and the saturated function SFx. By pressing the "CIRCUIT" button, the window in Fig **16** appears to select the active device used for synthesis process. Once selected the opamp model, the basic cell shown in chapter 4 is used to synthesize the current and voltage saturated function SFx. The parameters for that basic cell are evaluated by using (4.8) where the saturated function SFx parameters such as k, α, h and p=q are related to the opamp finite gain model. On the other hand, the parameters of the chaotic system in (1) are calculated by (2). Therefore, the user can generate an analog circuit based on opamps as shown in Fig. **17**.

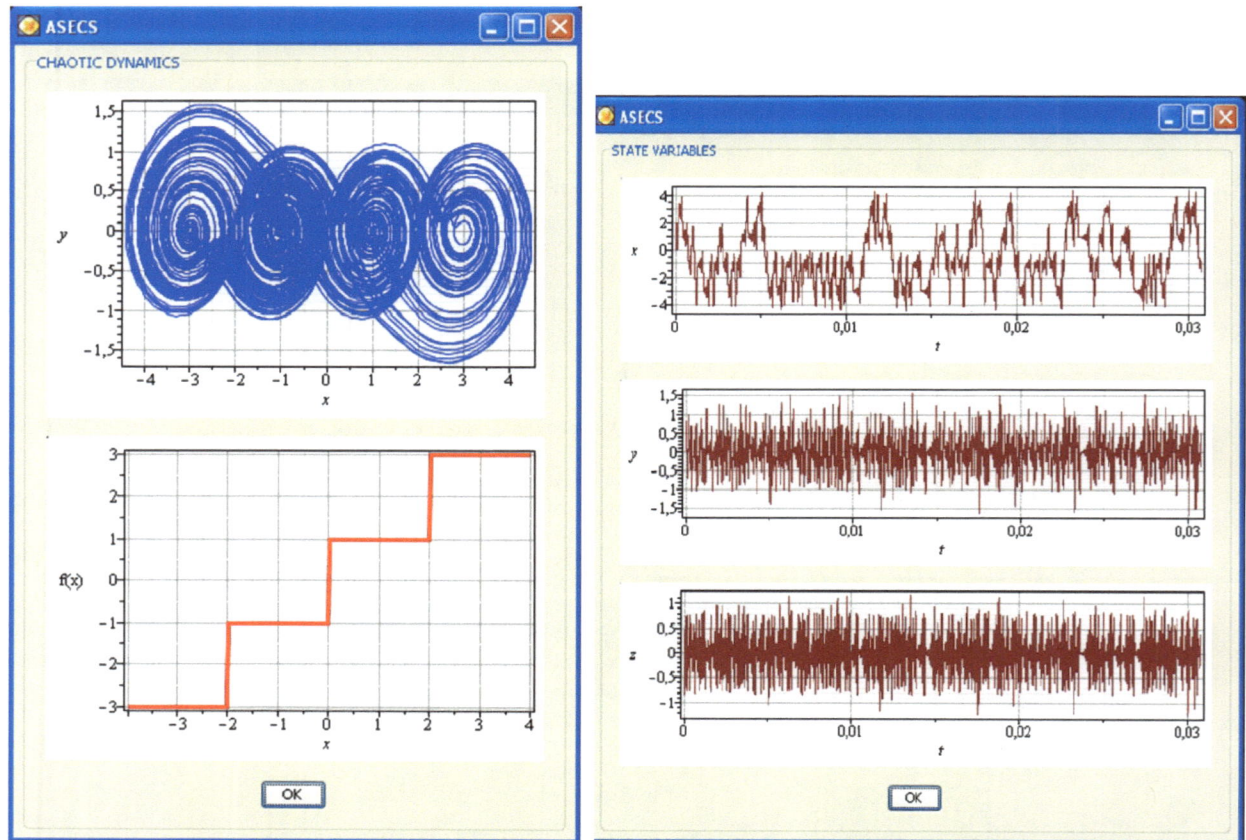

Figure 15: Screenshots of the nonlinear dynamics windows.

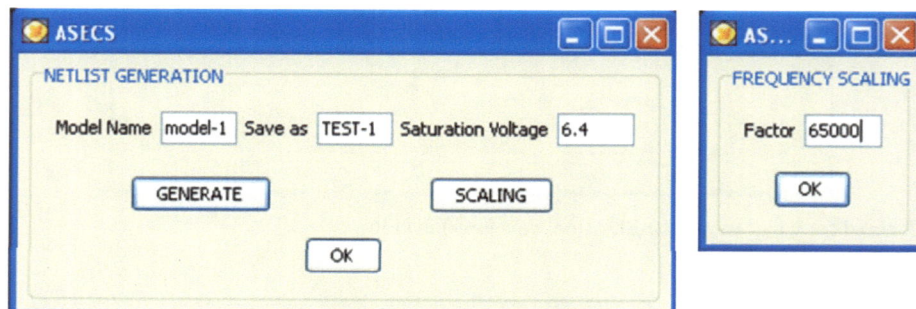

Figure 16: Screenshots for netlist generation window and frequency scaling window.

According to the system parameters obtained in the previous step (High-level simulation) and selecting Vsat= ±6.4V, the tool evaluates (4.8) and. (2) and it obtains: Rix=10KΩ, C=2.2nF, Rp=7KΩ, Rx=Ry=Rz=10KΩ, Rm=Rn=10KΩ, Rc=64KΩ, Ri=1KΩ, Rf=1MΩ, E1=±2V. The capacitor C is evaluated by the user frequency requirements according to the scaling factor of 65000 (f=10KHz). A netlist for the circuits shown in Fig. **17** is also generated by the tool and the users can execute a SPICE simulation as shown in Fig. **18**. In this manner, behavioural modelling is used during the design space exploration. After interactively using the tool to design multi-scrolls chaotic systems, the designers can verify and refine the resultant circuits using more accurate opamp models before the physical implementation. On the other hand, productivity is enhanced by allowing the user to focus only on the parameters of interest, shielding them from unnecessary complexity.

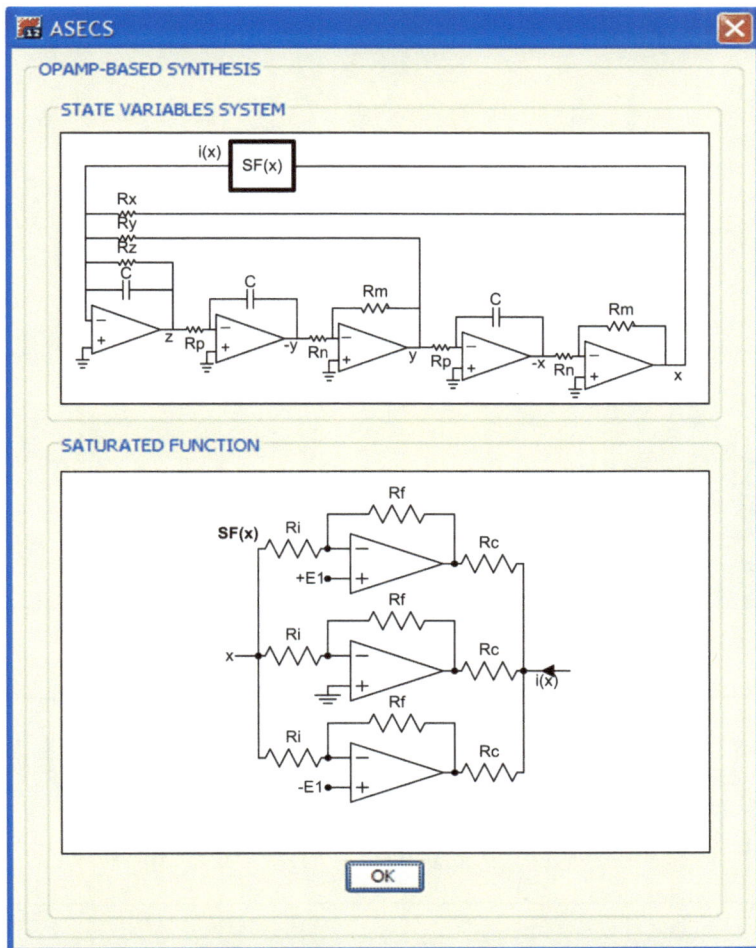

Figure 17: Screenshot of the opamp-based synthesis for 3D-6-scrolls chaotic attractor.

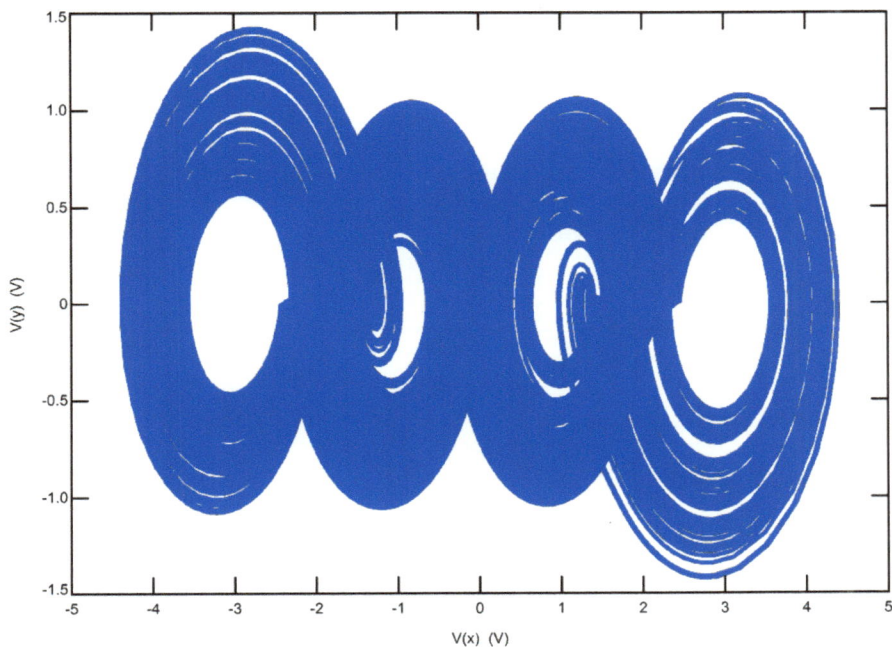

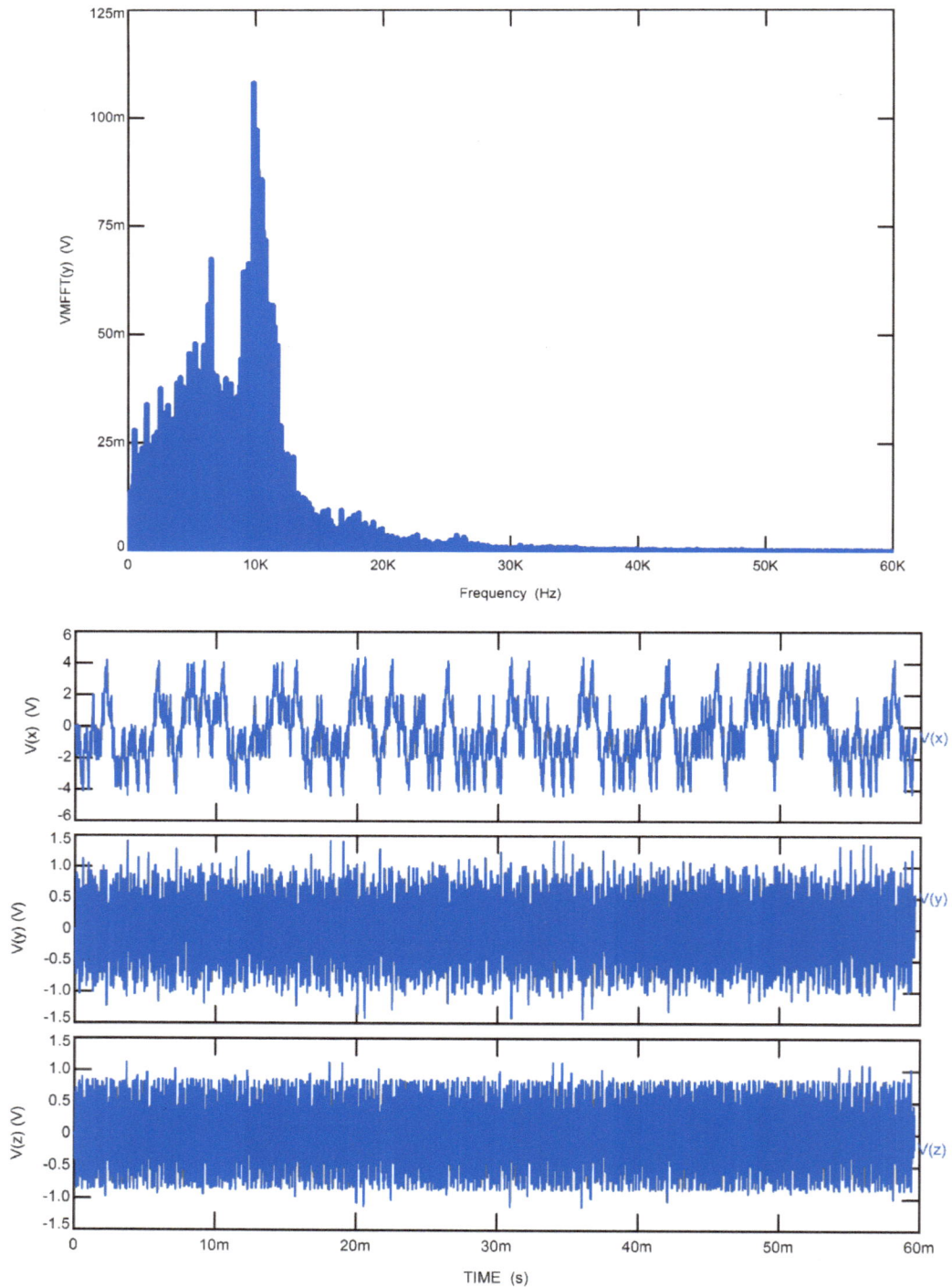

Figure 18: SPICE simulation results for case study.

EXPERIMENTAL REALIZATIONS

The synthesized circuit shown in Fig. **17** was physically implemented in [136]. The opamp TL081 [107] was used to physically realize the saturated function and state variables system. By selecting power supplies of Vdd = +7V and Vss = -7.8V, one obtains symmetrical saturation voltages of Vsat = ±6.4V. The experimental results are shown from Fig. **19** to Fig. **22**.

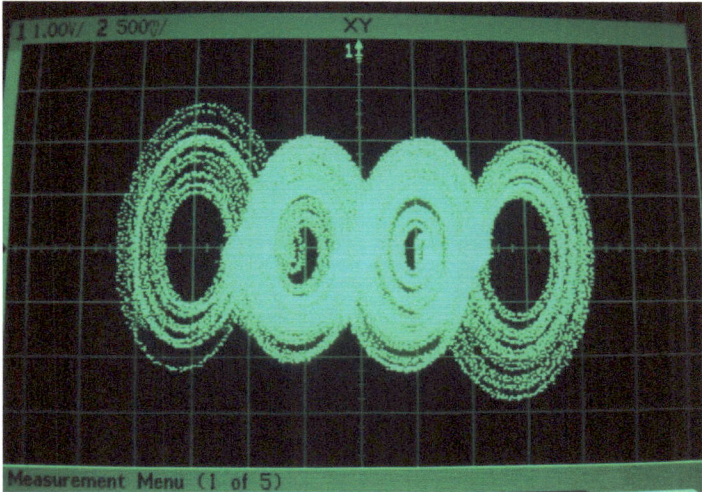

Figure 19: 1D-4-scrolls chaotic attractor (channel X=1V/div, channel Y=500mV/div).

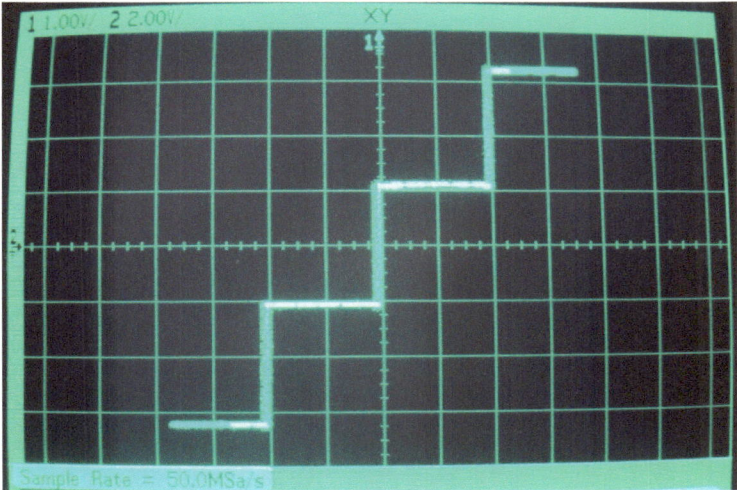

Figure 20: Saturated function (channel X=1V/div, channel Y=1V/div).

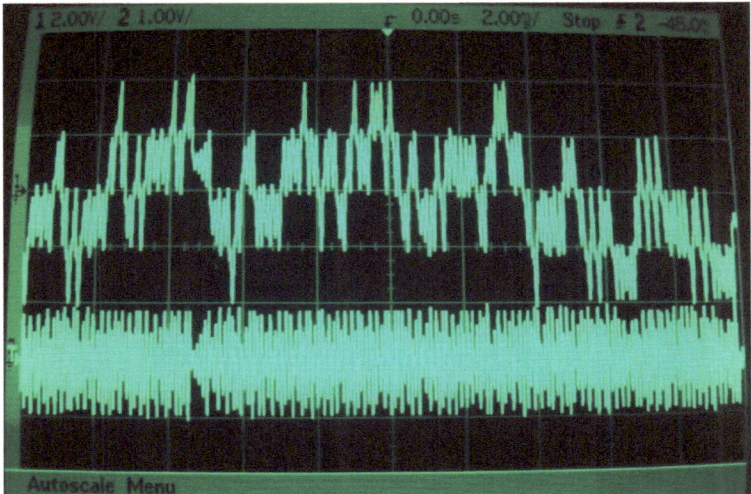

Figure 21: System variables (x and z) in time domain (channel 1=2V/div, channel 2=1V/div).

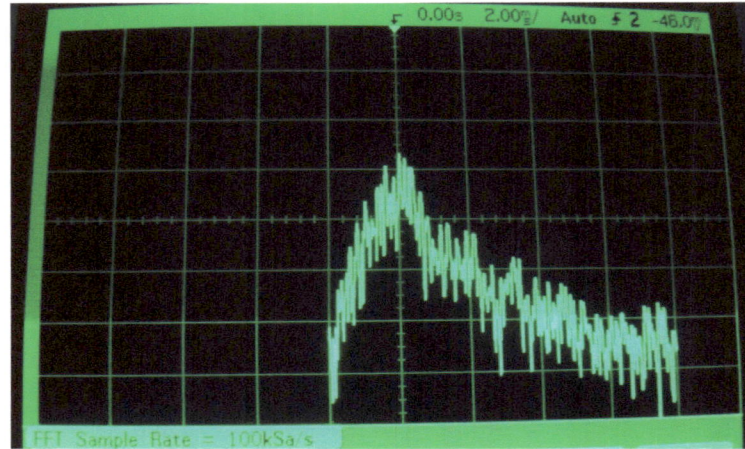

Figure 22: Frequency spectrum of the attractor (f=10 KHz).

The synthesis of 2D-multi-scrolls chaos generators can be found in [134]. For instance, the experimental realization of a 2D-3x3-scrolls attractor is shown in Fig. **23**. This attractor was synthesized with the following synthesis values: a=d1=0.7, b=c=d2=d3=0.7, k1=k2=1, h1=h2=1, p1=q1=p2=q2=1, α=6.4e-3, s=156.2 in (3.18) and (3.49). The experimental values are: Rix=10KΩ, C=2.2nf, Rp=Rm=Rn=7KΩ, Rx=Ry=Rz=10KΩ in Fig. **5** and Rix=10KΩ, Rc=64KΩ, Ri=1KΩ, Rf=1MΩ, R=10 KΩ, E1=±1V, Vsat=±6.4V, Vdd=+7.06V, Vss=-7.78V, and F=10KHz in Chapter 4 Fig. **8**(a). The opamp TL081 was used for all physical implementations.

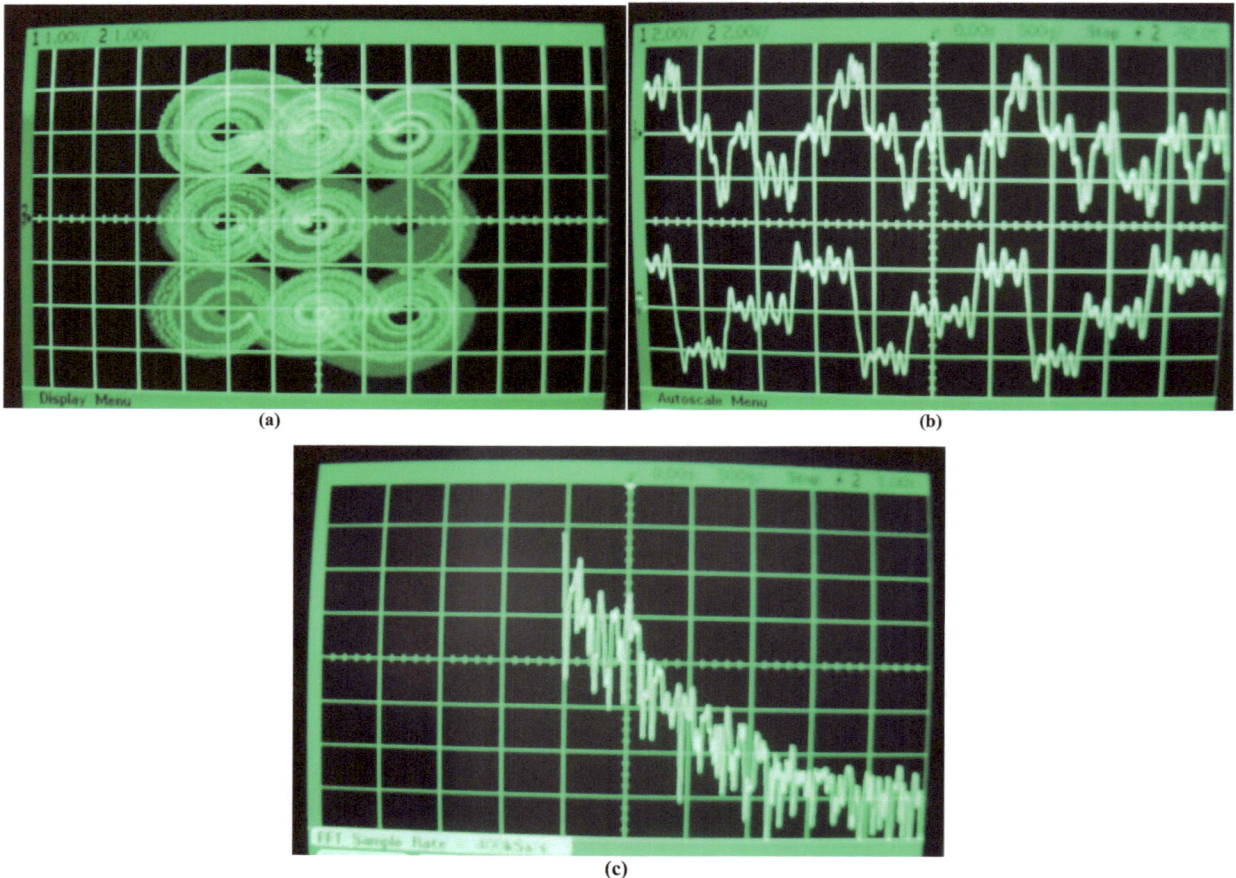

(a)

(b)

(c)

Figure 23: (a)2D-3-scrolls mesh chaotic attractor (channel X=1V/div, channel Y=1V/div), (b) System variables (x and y) in time domain (channel 1=2V/div, channel 2=2V/div) and (c) Frequency spectrum of the 2D-4-scrolls attractor (f=10 KHz).

The experimental realization of a 2D-4x4-scrolls attractor is shown in Fig. **24**. This attractor was synthesized with the following synthesis values: a=d1=0.7, b=c=d2=d3=0.7, k1=k2=1, h1=h2=2, p1=q1=p2=q2=1, α=6.4e-3, s=156.2 in (3.18) and (3.49). The experimental values are: Rix=10KΩ, C=2.2nf, Rp=Rm=Rn=7KΩ, Rx=Ry=Rz=10KΩ in Fig. **5** and Rix=10KΩ, Rc=64KΩ, Ri=1KΩ, Rf=1MΩ, R=10 KΩ, E1=±2V, Vsat=±6.4V, Vdd=+7.06V, Vss=-7.78V, and F=10KHz in Chapter 4 Fig. **8**(b).

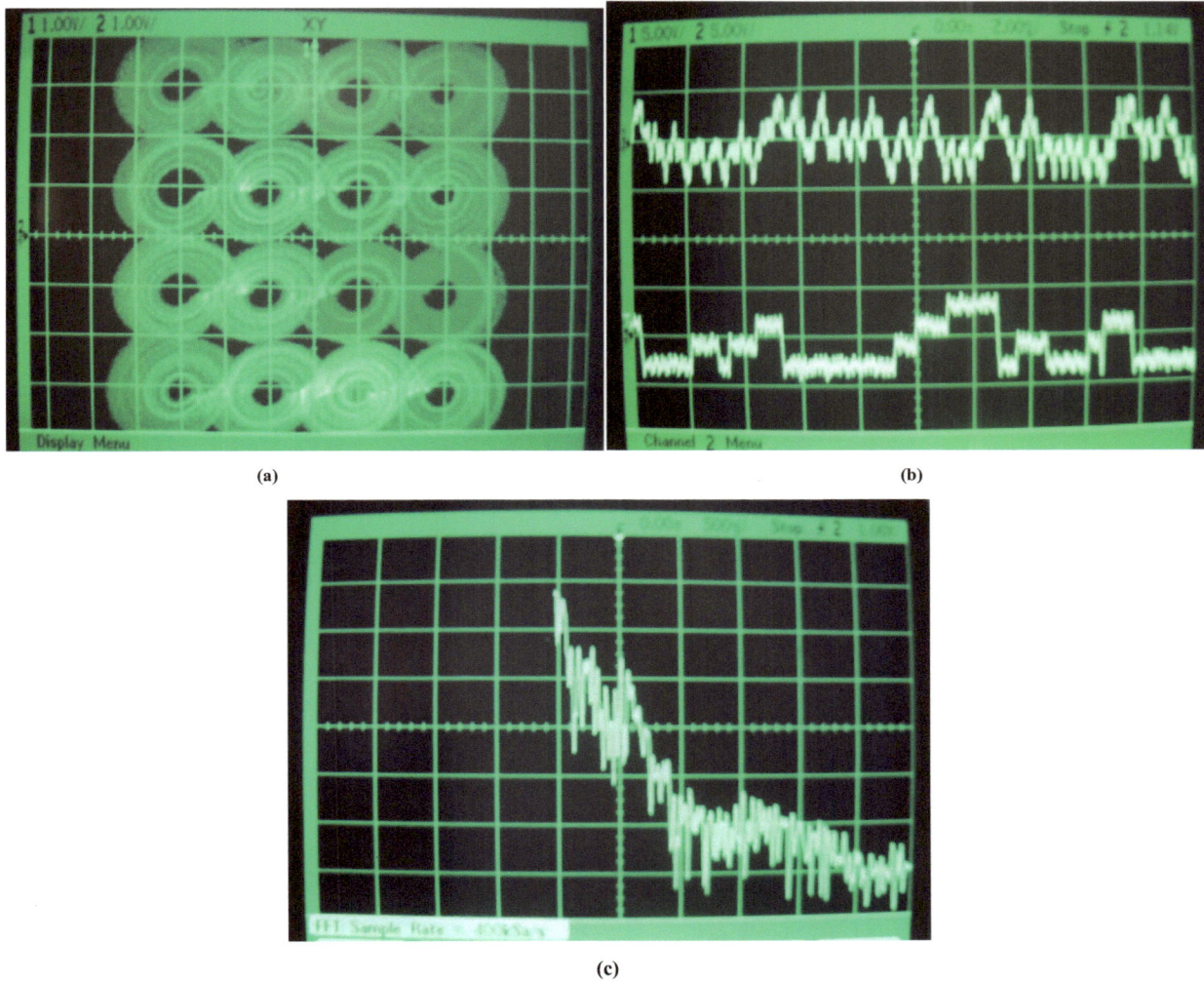

(a)

(b)

(c)

Figure 24: (a)2D-4-scrolls mesh chaotic attractor (channel X=1V/div, channel Y=1V/div), (b) System variables (x and y) in time domain (channel 1=5V/div, channel 2=5V/div) and (c) Frequency spectrum of the 2D-4-scrolls attractor (f=10 KHz).

APPLICATION TO SECURE COMMUNICATIONS

Chaos theory has been drawing a great deal of attention in the scientific community for almost two decades [21]. Remarkable research efforts have been invested in recent years, trying to export concepts from physics and mathematics into real-world engineering applications [105]. Recently, chaotic encryption to address the secure communication problem has received a great deal of attention [11,42]. Secure/private communication schemes using chaotic encryption are based on chaotic synchronization [9,10]. In this manner, the interest in the synchronization arises from the possibilities of encoding messages using as analog carriers the chaotic signal generated as a state, or as an output, of a chaotic system, called the transmitter. For the decoding process to be reliable, a second chaotic system, called the receiver, is proposed which is synchronized with the transmitter chaotic behaviour. The final detection stage simply consists in subtracting the transmitted signal from the locally generated synchronized state [9].

The aim of this section is to present a communication scheme to transmit encrypted information, which is based on the synchronization of multi-scroll chaotic attractors. The synchronization can be achieved by using Generalized Hamiltonian forms because it has principal advantages over other synchronization methods reported in the literature [10], like: (a) Enables synchronization to be achieved in a systematic way and clarifies the issue of deciding on the nature of the output signal to be transmitted. (b) It can be successfully applied to several well-known chaotic systems. (c) It does not require the computations of any Lyapunov exponent. (d) It does not require initial conditions belonging to the same basin of attraction. In the other hand, the communication scheme is based on chaotic additive masking to transmit analog information [31,105]. Theoretical results and SPICE simulations results are confirmed by experimental realizations.

Hamiltonian Forms and Observer Design for Multi-Scrolls Chaotic System

The multi-scroll chaotic system in (1) has been successfully synthesized by applying behavioural modelling as shown in the previous chapters. For sake of simplicity, (1) can be recast by (5) in order to change the name of the state variables with $x = x_1$, $y = x_2$, $z = x_3$.

$$\begin{bmatrix} \dfrac{dx_1}{dt} \\ \dfrac{dx_2}{dt} \\ \dfrac{dx_3}{dt} \end{bmatrix} = \begin{bmatrix} 0 & \dfrac{1}{RpC} & 0 \\ 0 & 0 & \dfrac{1}{RpC} \\ -\dfrac{1}{R_{x1}C} & -\dfrac{1}{R_{x2}C} & -\dfrac{1}{R_{x3}C} \end{bmatrix} \begin{bmatrix} x_1 \\ x_2 \\ x_3 \end{bmatrix} + \begin{bmatrix} 0 \\ 0 \\ \dfrac{i(x_1)Rix}{RixC} \end{bmatrix} \tag{5}$$

According to Hamiltonian synchronization theory [10], a general system of the form $\dot{x} = f(x)$, $x \in \Re^n$ can be rewritten in the Generalized Hamiltonian form as

$$\dot{x} = J(x)\frac{\partial H}{\partial x} + (I + S)\frac{\partial H}{\partial x} + F(y) \qquad x \in \Re^n \qquad y = C\frac{\partial H}{\partial x} \qquad y \in \Re^m \tag{6}$$

where the first term represents the conservative part of the system, and the second and third the non-conservative part of the system being F(y) the destabilizing part of it and having the relations

$$J(x) + J^T(x) = 0 \qquad S(x) = S^T(x) \tag{7}$$

The estimate of the state x(t) can be denoted by ξ(t), which considers the Hamiltonian energy function H(ξ) to be the particularization of H in terms of ξ(t) [9]. Similarly, we denote by η(t) the estimated output, computed in terms of the estimated state ξ(t). The gradient vector $\partial H(\xi)/\partial \xi$ is of the form Mξ with M being a constant, symmetric, positive definite matrix. A nonlinear state observer [10] for the Generalized Hamiltonian form (6) is given by

$$\dot{\xi} = J(y)\frac{\partial H}{\partial \xi} + (I + S)\frac{\partial H}{\partial \xi} + F(y) + K(y - \eta) \qquad \xi \in \Re^n \qquad \eta = C\frac{\partial H}{\partial \xi} \qquad \eta \in \Re^m \tag{8}$$

where K is the observer gain. The state estimation error, defined as e(t) = x(t) – ξ(t) and the output estimation error, defined as e_y(t) = y(t) – η(t), are governed by

$$\dot{e} = J(y)\frac{\partial H}{\partial e} + (I + S - KC)\frac{\partial H}{\partial e} \qquad e \in \Re^n \qquad e_y = C\frac{\partial H}{\partial e} \qquad e_y \in \Re^m \tag{9}$$

The slave system described by (8) synchronizes with the chaotic master system in Generalized Hamiltonian form (6), if $\lim_{t \to \infty} \|x(t) - \xi(t)\| = 0$.

No matter which initial conditions x(0) and ξ(0) have. The state estimation error e(t)=x(t)–ξ(t) represents the synchronization error. The master system of (5) can be expressed according to (6), by

$$
\begin{pmatrix} \dot{x}_1 \\ \dot{x}_2 \\ \dot{x}_3 \end{pmatrix} =
\begin{pmatrix} 0 & \dfrac{R_{x2}}{2Rp^2C} & \dfrac{1}{2RpC} \\[2mm] -\dfrac{R_{x2}}{2Rp^2C} & 0 & \dfrac{1}{RpC} \\[2mm] -\dfrac{1}{2RpC} & -\dfrac{1}{RpC} & 0 \end{pmatrix} \dfrac{\partial H}{\partial x}
+ \begin{pmatrix} 0 & \dfrac{R_{x2}}{2Rp^2C} & -\dfrac{1}{2RpC} \\[2mm] \dfrac{R_{x2}}{2Rp^2C} & 0 & 0 \\[2mm] -\dfrac{1}{2RpC} & 0 & -\dfrac{1}{R_{x3}C} \end{pmatrix} \dfrac{\partial H}{\partial x}
+ \begin{pmatrix} 0 \\ 0 \\ \dfrac{F(x_1)}{R_{ix}C} \end{pmatrix}
\tag{10}
$$

taking as Hamiltonian energy function the scalar function and gradient vector as

$$
H(x) = \frac{1}{2}\left[\frac{Rp}{R_{x1}}x_1^2 + \frac{Rp}{R_{x2}}x_2^2 + x_3^2 \right] \quad
\frac{\partial H}{\partial x} = \begin{pmatrix} \dfrac{Rp}{R_{x1}} & 0 & 0 \\[2mm] 0 & \dfrac{Rp}{R_{x2}} & 0 \\[2mm] 0 & 0 & 1 \end{pmatrix}
\begin{pmatrix} x_1 \\ x_2 \\ x_3 \end{pmatrix}
= \begin{pmatrix} \dfrac{Rp}{R_{x1}}x_1 \\[2mm] \dfrac{Rp}{R_{x2}}x_2 \\[2mm] x_3 \end{pmatrix}
\tag{11}
$$

It can be seen that $x1$ will be the coupling signal y which is transmitted from the master circuit to the observer. This means that $C = (0, 0, 1/RixC)$. The equation of the observer of (5) according to (8) is given by

$$
\begin{pmatrix} \dot{\xi}_1 \\ \dot{\xi}_2 \\ \dot{\xi}_3 \end{pmatrix} =
\begin{pmatrix} 0 & \dfrac{R_{x2}}{2Rp^2C} & \dfrac{1}{2RpC} \\[2mm] -\dfrac{R_{x2}}{2Rp^2C} & 0 & \dfrac{1}{RpC} \\[2mm] -\dfrac{1}{2RpC} & -\dfrac{1}{RpC} & 0 \end{pmatrix} \dfrac{\partial H}{\partial \xi}
+ \begin{pmatrix} 0 & \dfrac{R_{x2}}{2Rp^2C} & -\dfrac{1}{2RpC} \\[2mm] \dfrac{R_{x2}}{2Rp^2C} & 0 & 0 \\[2mm] -\dfrac{1}{2RpC} & 0 & -\dfrac{1}{R_{x3}C} \end{pmatrix} \dfrac{\partial H}{\partial \xi}
+ \begin{pmatrix} 0 \\ 0 \\ \dfrac{F(y)}{R_{ix}C} \end{pmatrix}
+ \begin{pmatrix} k_1 \\ k_2 \\ k_3 \end{pmatrix} e_y
\tag{12}
$$

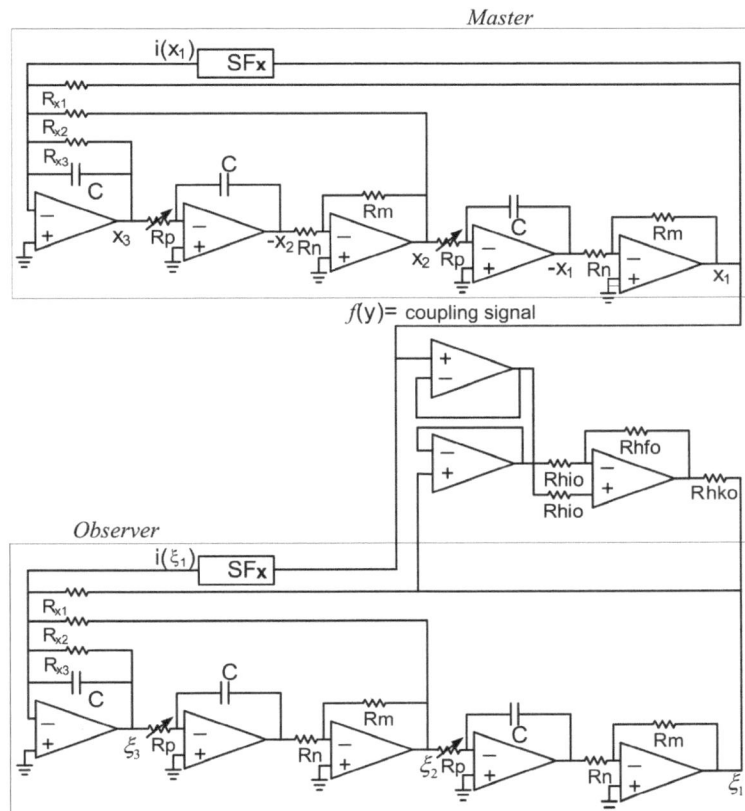

Figure 25: Proposed schematic diagram to synchronize multi-scrolls chaotic attractors.

As one can infer, the nonlinear component of the observer circuit is expected to be controlled directly by the master circuit. As a result, the proposed scheme to synchronize multi-scrolls attractors from (10) and (12) is shown in Fig. **25**. Vector k_n in (12) is the observer's gain and it is adjusted according to the sufficient conditions for synchronization [10]. By selecting Rix=10KΩ, C=2.2nf (Fo=10KHz), Rp=7KΩ, Rx1=Rx2=Rx3=10KΩ, Rm=Rn=10KΩ in (10) and (12), Rc=64KΩ, Rf=1MΩ, Ri=1KΩ, E1=±2V, Vsat=±6.4 for the saturated function SFx in Fig. **25**, and Rhio=10kΩ, Rhfo=3.9MΩ, Rhko=18Ω in Fig. **25**, one obtains the SPICE simulations shown in Fig. **26** to **28** that are in close accordance with the theoretical results. The coincidence of states is represented by a straight line, with slope equal to unity, in the phase plane for each state as shown in Fig. **28**. The opamp TL081 [107] was used to implement the chaotic attractors in Fig. **25**.

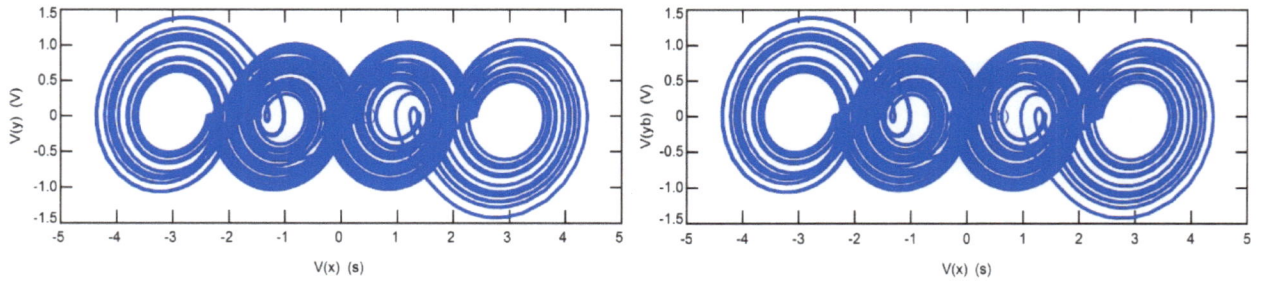

Figure 26: 1D-4-scroll chaotic attractors of the master circuit and the observer, respectively.

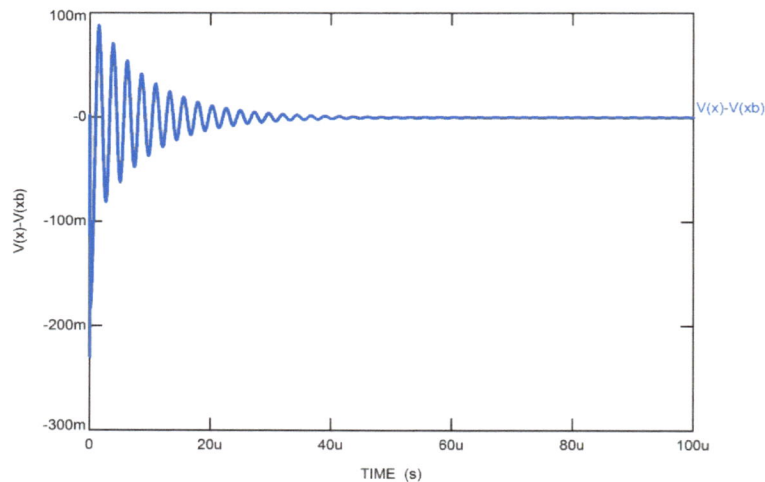

Figure 27: Error between the synchronized circuits in time domain.

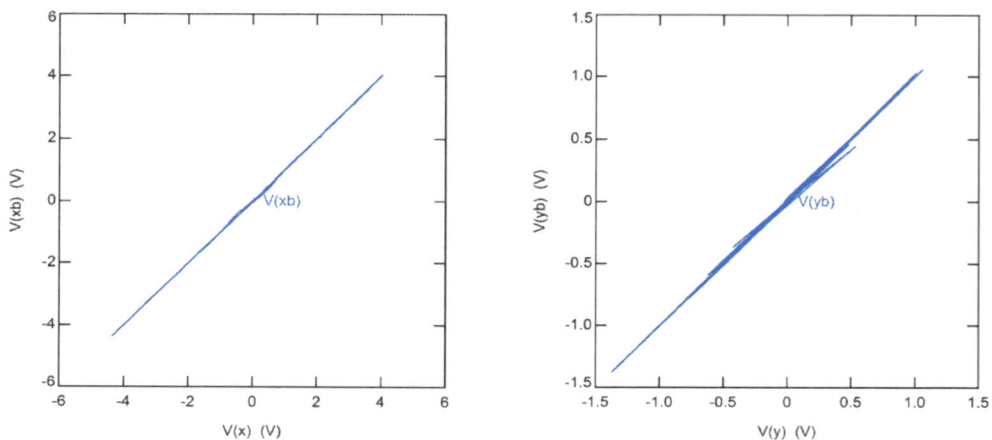

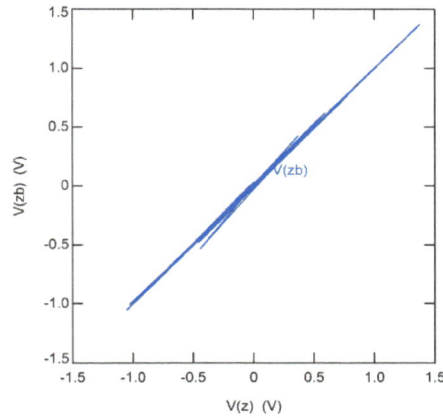

Figure 28: Phase plane diagrams for state variables in (10) and (12).

Secure Communication Scheme

The experimental set-up to transmit private analog signals by chaotic additive masking is shown in Fig. **29**. The basic idea of this scheme is as follows: first, one generates two multi-scrolls chaotic attractors from system in (5) and then, uses two transmission channels to avoid noise in the communication channel [105]. According to the observer equation (12), the system requires the injection of a current proportional to the error signal, between state $x1$ and $\xi1$. Therefore, the two chaotic systems are synchronized when the chaotic signal $x1(t)$ is transmitted by the first channel; while the confidential message $m(t)$ is encrypted in the chaotic signal $x2(t)$ by means of a additive process as shown in Fig. **29**. In this manner, the confidential information is send by the second channel. To recover the original message, it is only necessary to apply the reverse operation.

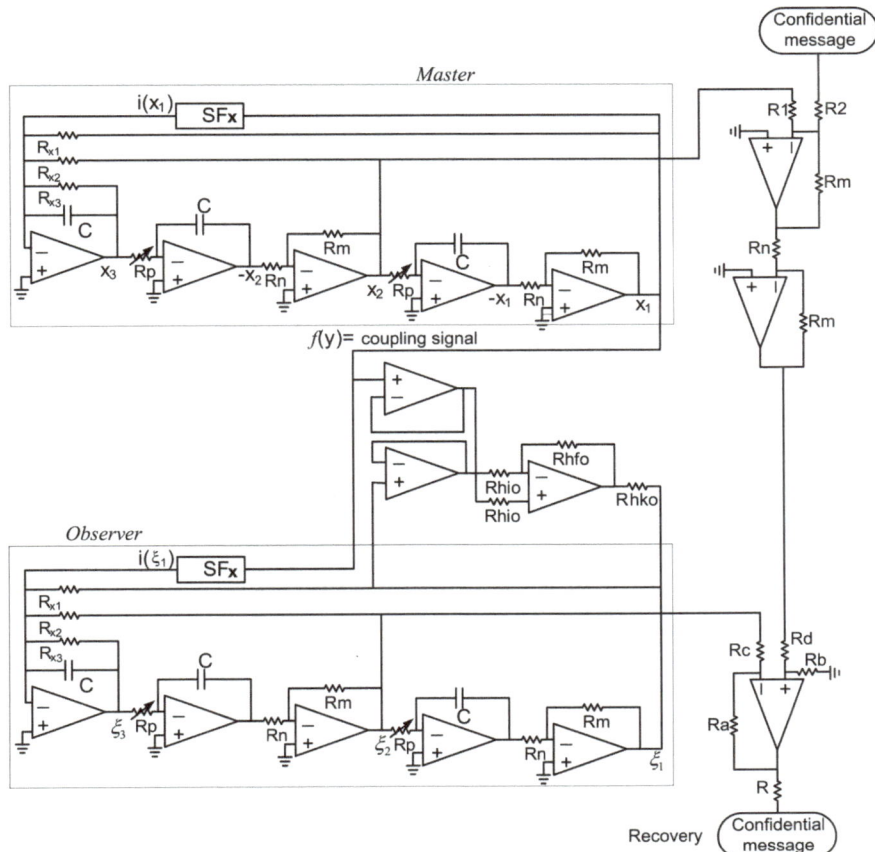

Figure 29: Transmission of secure information by applying chaotic additive masking scheme.

Experimental Realization

In Fig. **30** is shown the 1D-4-scrolls chaotic attractors that generate the master and observer systems. The synchronization of the states is shown in Fig. **31**. Finally, in order to demonstrate the transmission of confidential information, the scheme in Fig. **29** was used to transmit an audio signal as shown in Fig. **32** where is shown the confidential message *m(t)*, the chaotic signal transmitted by the public channel *x2(t)+m(t)* and the recovered signal in the receptor $\hat{m}(t)$.

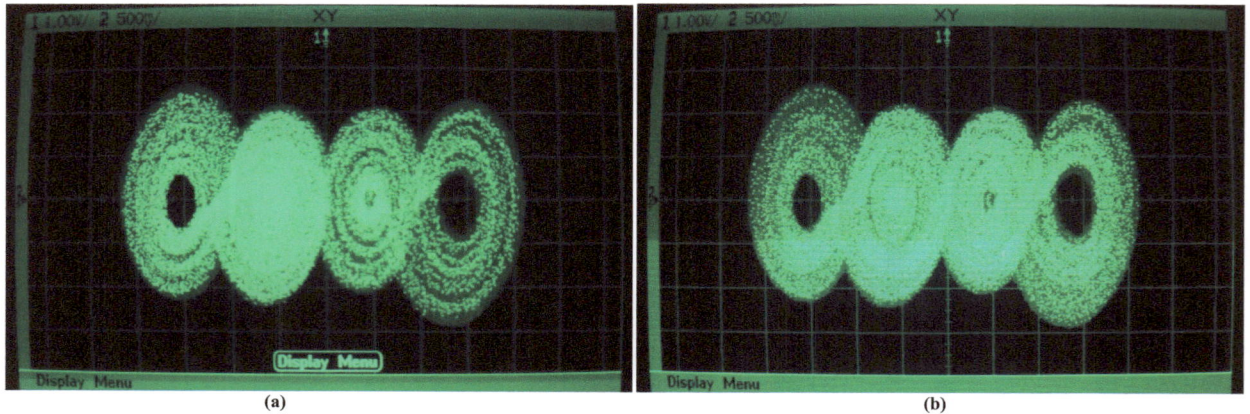

Figure 30: 1D-4-scrolls chaotic attractors for: (a) master circuit (x=1V/div, y=500mV/div) and (b) slave circuit (x=1V/div, y=500mV/div).

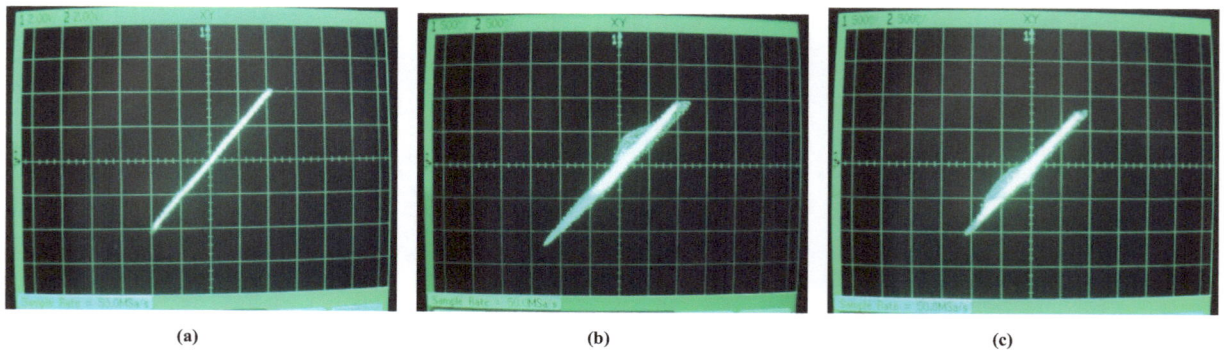

Figure 31: Phase planes of the sates: (a) x1 vs ξ1, (b) x2 vs ξ2 and (c) x3 vs ξ3.

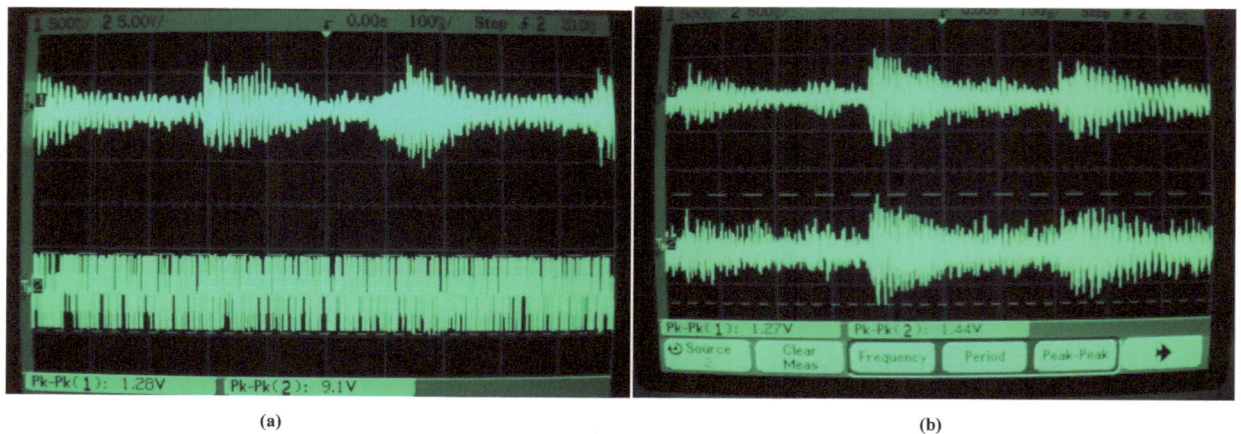

Figure 32: Encrypted communication of audio signals using chaotic additive masking scheme: (a) ch1: *m(t)*, ch2:*x2(t)+m(t)*; (b) ch1: *m(t)*, ch2: $\hat{m}(t)$.

CHAPTER 6

General Conclusion and Future Work

Abstract: In this final chapter, the results of the previous chapters are recapitulated and some lines for future research are provided.

Keywords: Chaos, electronic design automation, dynamical systems, chaos generators, modelling and simulation.

GENERAL CONCLUSION

This book has defined, developed and demonstrated how to implement an automatic synthesis methodology for the design of multi-scrolls chaotic systems with multiple orientations (1D, 2D and 3D). First, it was described that the modelling of nonlinear functions by PWL characteristics to allow the formulation of state variables systems in order to model the behaviour of three kinds of chaotic oscillators (Chua's circuit, Generalized Chua's circuit and multi-scrolls chaotic oscillator). To explore its dynamics, numerical simulation was executed. Furthermore, it has been shown the appropriateness of implementing automatic control and determination of step-size in multistep Adams-Moulton algorithm to speed-up time simulation of nonlinear dynamical systems such as chaotic oscillators.

It was also shown that the implementation of procedures for scaling the excursion level and frequency of chaotic signals leads to enhanced automatic design tasks by controlling the breaking points and slopes. Consequently, it allows us to synthesize the PWL functions with real opamps.

As a result, the behavioural modelling can be used to explore the chaotic dynamics and to estimate the values of the circuit elements before the physical implementation of electronic circuits.

A simple algorithm to compute Lyapunov exponents of multi-dimensional nonlinear dynamical systems, such as the PWL-based chaotic oscillators was introduced. The numerical method is based on the solution of m-PW variational systems and uses the standard Gram-Schmidt orthogonalization process. Due to this, the algorithm could be used in the synthesis approach where several solutions must be evaluated in order to determine if the potential solution presents chaotic behaviour.

In addition, the synthesis of current and voltage saturated function series using practical opamps was oriented to implement the PWL approximations by controlling their breakpoints and slopes. In particular, this book introduced an opamp based-basic cell and two structures for connecting this basic cell. In addition, a Verilog-A model was used to include non-ideal effects for opamps and to make the simulation time less critical in contrast to when using complex SPICE models. In this manner, the synthesis of 1D 3- and 6-scrolls chaotic attractors, 2D 3- and 4-scrolls chaotic attractors and 3D 3- and 4-scrolls chaotic attractors using opamps, was presented. Consequently, a new Electronic Design Automation (EDA)-tool based on behavioural modelling for the design of multi-scrolls chaotic systems has been introduced. The guidelines to implement this tool were shown, which can provide with a GUI to facilitate design process avoiding unnecessary complexity.

The validity of the synthesized multi-scroll chaos generator through its physical implementation using commercially available opamps, was demonstrated. To highlight the usefulness of the multi-scroll chaotic systems, two 1D-4-scroll chaos generators were synchronized to implement a secure communication system by applying Hamiltonian forms and observer approach. Other multi-directional chaos generators can be used to implement this kind of secure communication systems. In such a case, 2D or 3D multi-scroll chaos generators can be synthesized with the proposed EDA tool provided in Chapter 5.

Finally, since SPICE simulations and experimental realization are in good agreement with the ESL numerical simulations, one can conclude on the usefulness of this automatic synthesis methodology based on high-level behavioural modelling to synthesize 1D, 2D and 3D-multi-scrolls chaotic systems. In this manner, this book introduces a contribution on the development of an ADA tool for nonlinear dynamical systems, specifically for multi-scrolls chaotic oscillators based on saturated nonlinear function series with multi-segments.

Jesus Manuel Muñoz Pacheco and Esteban Tlelo Cuautle

FUTURE WORK

In electronics, analog circuit design typically needs to take into account several performance specifications, which depend on the circuit designer's abilities to successfully exploit a range of nonlinear behaviours across different levels of abstraction from devices to circuits and systems. Managing all these aspects is what makes analog circuit design cumbersome and challenging. Therefore, efficient synthesis methodologies supported by appropriate analog EDA programs need to be developed. On the other hand, the integrated circuit (IC) design effort has been continuously increasing with the integration of more and more functionalities onto a single chip. For instance, in the area of CMOS analog IC design, increasing design complexity is widening the gap between the complex systems and the ability to design them. From this point of view, there are many opportunities for future work that spring from this book. For example, no real attention has been given to the methods used to determine the chaotic regimen. One can search for the development of methods to compute Lyapunov exponents and bifurcation diagrams to find systematic design strategies for the representation of the chaotic behaviour with respect to a particular parameter of the chaotic system. Along with this problem, one can search for the discovering of the range of values for that particular parameter where the dynamical system exhibits chaotic behaviour.

Last but not least, an extension to the synthesis methodology could be investigated in order to consider another kind of electronic devices such as current feedback operational amplifiers (CFOAs) and Current Conveyors for the synthesis process of chaotic systems. The implementation of multi-scroll chaos generators with IC technology is another problem when using sub-micrometric CMOS technology.

All this open problems can be included in the search for new procedures to translate from the electronic system level (ESL) down to block specifications (opamp-level) and down to IC transistor level in order to design integrated chaotic systems.

The application of multi-dimensional multi-scroll chaos generators is another open problem. For instance, one can realize secure communication systems as the ones shown in Chapter 5.

APPENDIX

Macromodels

TL081 OPERATIONAL AMPLIFIER "MACROMODEL" SUBCIRCUIT

```
*CONNECTIONS:
*                 NON-INVERTING INPUT
*                 | INVERTING INPUT
*                 | | POSITIVE POWER SUPPLY
*                 | | | NEGATIVE POWER SUPPLY
*                 | | | | OUTPUT

.SUBCKT TL081    1 2 3 4 5
 C1   11 12 3.498E-12
 C2    6  7 15.00E-12
 DC    5 53 DX
 DE   54  5 DX
 DLP  90 91 DX
 DLN  92 90 DX
 DP    4  3 DX
 EGND 99  0 POLY(2) (3,0) (4,0) 0 .5 .5
 FB    7 99 POLY(5) VB VC VE VLP VLN 0 4.715E6 -5E6 5E6 5E6 -5E6
 GA    6  0 11 12 282.8E-6
 GCM   0  6 10 99 8.942E-9
 ISS   3 10 DC 195.0E-6
 HLIM 90  0 VLIM 1K
 J1   11  2 10 JX
 J2   12  1 10 JX
 R2    6  9 100.0E3
 RD1   4 11 3.536E3
 RD2   4 12 3.536E3
 RO1   8  5 150
 RO2   7 99 150
 RP    3  4 2.143E3
 RSS  10 99 1.026E6
 VB    9  0 DC 0
 VC    3 53 DC 2.200
 VE   54  4 DC 2.200
 VLIM  7  8 DC 0
 VLP  91  0 DC 25
 VLN   0 92 DC 25
.MODEL DX D(IS=800.0E-18)
.MODEL JX PJF(IS=15.00E-12 BETA=270.1E-6 VTO=-1)
.ENDS
```

TLC2262 OPERATIONAL AMPLIFIER "MACROMODEL" SUBCIRCUIT

```
*CONNECTIONS:
*                 NON-INVERTING INPUT
*                 | INVERTING INPUT
*                 | | POSITIVE POWER SUPPLY
*                 | | | NEGATIVE POWER SUPPLY
*                 | | | | OUTPUT
*                 | | | | |
```

```
.SUBCKT  TLC2262   1 2 3 4 5
 C1   11 12 3.073E-12
 C2    6  7 15.00E-12
 DC    5 53 DX
 DE   54  5 DX
 DLP  90 91 DX
 DLN  92 90 DX
 DP    4  3 DX
 EGND 99  0 POLY(2) (3,0) (4,0) 0 .5 .5
 FB 7 99 POLY(5) VB VC VE VLP VLN 0 16.37E6 -20E6 20E6 20E6 -20E6
 GA    6  0 11 12 51.84E-6
 GCM 0  6 10 99 4.12E-9
 ISS   3 10 DC 8.250E-6
 HLIM 90  0 VLIM 1K
 J1   11  2 10 JX
 J2   12  1 10 JX
 R2    6  9 100.0E3
 RD1 60 11 19.29E3
 RD2 60 12 19.29E3
 RO1   8  5 110
 RO2   7 99 110
 RP    3  4 47.06E3
 RSS  10 99 24.24E6
 VAD  60 4 -.6
 VB    9  0 DC 0
 VC 3 53 DC .65
 VE   54  4 DC .65
 VLIM  7  8 DC 0
 VLP  91  0 DC .1
 VLN  0 92 DC 9.4
.MODEL DX D(IS=800.0E-18)
.MODEL JX PJF(IS=500.0E-15 BETA=651.4E-6 VTO=-.015)
.ENDS
```

REFERENCES

[1]   V. Grimblatt, "Synthesis - State of art", in *Proc. 6th International Caribbean Conference on Devices, Circuits, and Systems*, 2006, pp. 327-332.
[2]   W.M.C. Sansen; *Analog Design Essentials*, Dordrecht, NL: Springer, 2006.
[3]   F. Maloberti; *Analog Design for CMOS VLSI Systems,* Dordrecht, NL: Kluwer Academic, 2001.
[4]   Y.M. Li, S.M. Yu, Y.L. Li, "Electronic design automation using a unified optimization framework", *Mathematics and Computers in Simulation*, vol. 79, no. 4, pp. 1137-1152, 2008.
[5]   R. Castro-López, F.V. Fernández, O. Guerra-Vinuesa, Á. Rodríguez-Vázquez; *Reuse-Based Methodologies and Tools in the Design of Analog and Mixed-Signal Integrated Circuits,* Dordrecht, NL: Springer, 2006.
[6]   G. Gielen, "Design methodology and model generation for complex analog blocks", in *Proc. 14th Workshop on Advances in Analog Circuit Design*, 2006, pp. 113-141.
[7]   E. Martens, G. Gielen, "Classification of analog synthesis tools based on their architecture selection mechanisms", *Integration-the VLSI Journal*, vol. 41, issue: 2, pp. 238-252, 2008.
[8]   G. Van der Plas, G. Gielen, W.M.C. Sansen, *A Computer-Aided Design and Synthesis Environment for Analog Integrated Circuits*, Dordrecht, NL: Kluwer Academic, 2002.
[9]   A. Pikovsky, M. Rosenblum, J. Kurths, *Synchronization: A universal concept in nonlinear sciences*, Cambridge UK: Cambridge University Press, 2008.
[10]   H. Sira-Ramírez, C. Cruz-Hernández, "Synchronization of chaotic systems: A generalized Hamiltonian systems approach", *Int. J. Bifurcat. Chaos*, vol. 11, no. 5, pp. 1381-1395, 2001.
[11]   L. Gamez-Guzman, C. Cruz-Hernandez, R.M. Lopez-Gutierrez, "Synchronization of multi-scroll chaos generators: application to private communication", *Revista Mexicana de Fisica*, vol. 54, no. 4, pp. 299-305, 2008.
[12]   C. W. Wu, *Synchronization in coupled chaotic circuits and systems*, Singapore: World Scientific, 2002.
[13]   J.M. González-Miranda, *Synchronization and control of chaos*, Singapore: Imperial College Press, 2004.
[14]   T.L. Carroll and L. Pecora, "Synchronizing chaotic circuits", *IEEE Trans. Circuits Syst. I*, vol. 38, no. 4, 453-456, 1991.
[15]   S. Boccaletti, J. Kurths, G. Osipov, D.L. Valladares, C.S. Zhou, "The synchronization of chaotic systems", *Physics Reports*, vol. 366, pp. 1–101, 2002.
[16]   A.C.J. Lou, "A theory for synchronization of dynamical systems", *Communications in Nonlinear Science and Numerical Simulation*, vol. 14, issue 5, pp. 1901-1951, 2009.
[17]   G. Chen and T. Ueta, *Chaos in circuits and systems*, Singapore: World Scientific, 2002.
[18]   M. J. Ogorzalek, *Chaos and complexity in nonlinear electronic circuits*, Singapore: World Scientific, 1997.
[19]   Ji-Huan He, "Nonlinear science as a fluctuating research frontier", *Chaos Solit. Fract.*, vol. 41, issue 5, pp. 2533-2537, 2009.
[20]   J. Lu, G. Chen, "Generating multiscroll chaotic attractors: theories, methods and applications", *Int. J. Bifurcat. Chaos*, vol. 16, no. 4, pp. 775-858, 2006.
[21]   S.E. El-Khamy, "New trends in wireless multimedia communications based on chaos and fractals", in *Proc. Twenty-First National Radio Science Conference*, 2004, pp. 1-25.
[22]   J. Cordova-Zecena, "Chaotic Dynamical Systems and Their Applications", in *Proc. XXV CURCCAF*, 2001, pp. 167.
[23]   S. H. Strogatz; *Nonlinear Dynamics and Chaos: with Applications to Physics, Biology, Chemistry, and Engineering*, Boulder, CO: Westview Press, 2001.
[24]   P. A. Cook; *Nonlinear Dynamical Systems*, Hertfordshire, UK: Prentice Hall; 1994.
[25]   E. Ott, *Chaos in Dynamical systems*, Cambridge UK: Cambridge University Press, 1994.
[26]   P.Collet, J.P. Eckmann, *Concepts and Results in Chaotic Dynamics: A Short Course*, Heidelberg, DE: Springer-Verlag GmbH, 2006.
[27]   V.S. Anishchenko, V. Astakhov, A. Neiman, T. Vadivasova, L. Schimansky-Geier, *Nonlinear Dynamics of Chaotic and Stochastic Systems: Tutorial and Modern Developments*, Heidelberg, DE: Springer-Verlag GmbH, 2007.
[28]   E. Tlelo-Cuautle, A. Gaona-Hernández, J. García-Delgado; "Implementation of a chaotic oscillator by designing Chua's diode with CMOS CFOAs", *Analog Integrated Circuits and Signal Processing*; vol. 48, no. 2, pp. 159–162, 2006.

[29] J.M. Cruz, L.O. Chua, "A CMOS IC nonlinear resistor for Chua's circuit", *IEEE Trans. Circuits Syst. I*, vol. 39, pp. 985–995, 1992.

[30] C. Sánchez-López, A. Castro-Hernández and A. Pérez-Trejo, "Experimental verification of the Chua's circuit designed with UGCs", *IEICE Electron. Express*, vol. 5, no. 17, pp. 657-661, 2008.

[31] R. Trejo-Guerra, E. Tlelo-Cuautle, C. Cruz-Hernandez, C. Sanchez-Lopez, "Chaotic communication system using Chua's oscillators realized with CCII+s", *Int. J. Bifurcat. Chaos*, vol. 19, no. 12, pp. 4217-4226, 2009.

[32] A. Gálvez, "Numerical-Symbolic MATLAB Program for the analysis of three-dimensional chaotic systems", *Lecture Notes in Computer Science*, vol. 4488, pp. 211-218, 2007.

[33] L.G. Birta, G. Arbez; *Modelling and Simulation: Exploring Dynamic System Behaviour*; London, UK: Springer-Verlag; 2007.

[34] T.S. Parker, L.O. Chua; *Practical Numerical Algorithms for Chaotic Systems*; New York, USA: Springer-Verlag, 1992.

[35] F. C. Hoppensteadt; *Analysis and Simulation of Chaotic Systems*; New York, USA: Springer-Verlag; 2000.

[36] Zelinka, G.R. Chen, S. Celikovsky; "Chaos synthesis by means of evolutionary algorithms"; *Int. J. Bifurcat. Chaos*, vol. 18, no. 4, pp. 911-942, 2008.

[37] J. Lü, K. Muralib, S. Sinhac, H. Leungd and M.A. Aziz-Alaouie, "Generating multi-scroll chaotic attractors by thresholding", *Physics Letters A*, vol. 372, no. 18, pp. 3234-3239, 2008.

[38] F.Q. Wang, C.X. Liu, "Generation of multi-scroll chaotic attractors via the saw-tooth function", *International Journal of Modern Physics B*, vol. 22, no. 15, pp. 2399-2405, 2008.

[39] W. Liu, W.K.S. Tang, G. Chen, "2x2-scroll attractors generated in a three-dimensional smooth autonomous system", *Int. J. Bifurcat. Chaos*, vol. 17, pp. 4153-4157, 2007.

[40] C. Liu, T. Liu, L. Liu, Y. Su, "A New Nonlinear Chaotic System", *Int. J. Nonlinear Sci.*, vol. 7, no.3, pp. 345-352, 2006.

[41] G.Y. Qi, G.R. Chen, Y.H. Zhang, "On a new asymmetric chaotic system", *Chaos Solit. Fract.*, vol. 37, no. 2, pp. 409-423, 2008.

[42] H. Chuan-Kuei, T. Shuh-Chuan and W. Yu-Re, "Implementation of chaotic secure communication systems based on OPA circuits", *Chaos Solit. Fract.*, vol. 23, no. 2, pp. 589-600, 2005.

[43] W.P. Torres, A.V. Oppenheim, R.R. Rosales, "Generalized frequency modulation", *IEEE Trans. Circuits Syst. I*, vol. 48, no. 12, pp. 1405-1412, 2001.

[44] J. Yao, A.J. Lawrance, "Bit error rate calculation for multi-user coherent chaos-shift-keying communication systems", *IEICE Trans. Fund.*, vol. E87A, issue: 9, pp. 2280-2291, 2004.

[45] J.C. Feng, C.K. Tse, "Identification and tracking of chaotic signals with application to non-coherent detection for chaos-based communications", in *Proc. International Conference on Communication, Circuits, and Systems*, 2004, pp. 851-854.

[46] J. Lao, Q.S. Ren, J.Y. Zhao, "A Novel Chaotic Stream DS-UWB System", in *Proc. IEEE International Joint Conference on Neural Networks (IJCNN)*; 2008, pp. 835-839.

[47] H. Li, Y.L. Song, Y.W. Yang, "Analysis of multi-user chaotic pulse position modulation UWB system", in *Proc. Information Technology and Environmental System Sciences*, 2008, pp. 403-405.

[48] J. Ming, J. Cheng, G.X. Li, "Chaotic spread-spectrum sequences using chaotic quantization", in *Proc. International Symposium on Intelligent Signal Processing and Communication Systems*, 2007, pp. 48-51.

[49] S. Penaud, P. Bouysse, J. Guittard, R. Quere, A. Duverdier, "BER improvement of an asynchronous DS-CDMA system using chaotic spreading sequences", *Annals of Telecommunications*; vol. 58, pp. 656-672, 2003.

[50] P.P. Yupapin , S. Chaiyasoonthorn, S. Thongmee; "Chaotic Signals Filtering Device using the Series Waveguide Micro-Ring Resonators", in *Proc. International Workshop and Conference on Photonics and Nanotechnology*, 2008, pp. 42-47.

[51] J.C. Feng, H.J. Fan, C.K. Tse, "Convergence analysis of the unscented Kalman filter for filtering noisy chaotic signals", in *Proc. IEEE Int. Symp. Circuits and Systems*, 2007, pp. 1681-1684.

[52] B. Romeira, J.M.L. Figueiredo, T.J. Slight, L. Wang, E. Wasige, C.N. Ironside, J.M. Quintana, M.J. Avedillo, "Synchronization and chaos in a laser diode driven by a resonant tunnelling diode", *IET Optoelectronics*, vol. 2, issue: 6, pp. 211-215, 2008.

[53] J.M. Liu, H.F. Chen, S. Tang, "Optical-communication systems based on chaos in semiconductor lasers"; *IEEE Trans. Circuits Syst. I*, vol. 48, issue: 12, pp. 1475-1483, 2001.

[54] H. Dedieu, M.P. Kennedy and M. Hasler, "Chaotic shift keying: Modulation and demodulation of a chaotic carrier using self-synchronizing Chua's circuits". *IEEE Trans. Circuits Syst. II*, vol. 40, issue: 10, pp. 634-642, 1993.

[55] T. Yang and L.O. Chua, "Secure communication via chaotic parameter modulation", *IEEE Trans. Circuits Syst. I*, vol. 43, issue: 9, pp. 817-819, 1996.

[56] B. Mensour, A. Longtin, "Synchronization of delay-differential equations with application to private communication", *Phys Lett A*, vol. A244, issue: 1, pp. 59-70, 1998.

[57] L. Gámez-Guzmán, C. Cruz-Hernández, R.M. López-Gutiérrez y E.E. García-Guerrero, "Synchronization of Chua's circuits with multi-scroll attractors: Application to communication", *Communications in Nonlinear Science and Numerical Simulation*, vol. 14, pp. 2765-2775, 2009.

[58] C. Cruz-Hernández, A.A. Martynyuk. *Advances in chaotic dynamics with applications*, London, UK: Cambridge Scientific Publishers, 2009.

[59] L.P. Fang, H. Zhang and Q.Y. Tong, "Chaotic circuit, information and ordered space", *Int. J. Nonlinear Sci.*, vol. 8, issue: 1, pp. 59-62, 2007.

[60] J.A.K. Suykens, and J. Vandewalle, "Quasilinear approach to nonlinear systems and the design of n-double scroll (n = 1, 2, 3, 4, . . .)", *IEE Proc. G*, vol. 138, pp. 595-603, 1991.

[61] J.A.K. Suykens, A. Huang and L.O. Chua, "A family of n-scroll attractors from a generalized Chua's circuit", *Int. J. Electron. Commun.*, vol. 51, pp.131-138, 1997.

[62] M.A. Aziz-Alaoui, "Differential equations with multispiral attractors", *Int. J. Bifurcat. Chaos,* vol. 9, pp. 1009-1039, 1999.

[63] P. Arena, S. Baglio, L. Fortuna, and G. Manganaro, "Generation of n-double scrolls via cellular neural networks", *Int. J. Circuit Th. Appl.*, vol. 24, pp. 241-252, 1996.

[64] P. Arena, S. Baglio, L. Fortuna, and G. Manganaro, "State controlled CNN: A new strategy for generating high complex dynamics", *IEICE Trans. Fund.*, vol. E79-A, pp. 1647-1657, 1996.

[65] M.E. Yalcin, J.A.K. Suykens, and J. Vandewalle, "Experimental confirmation of 3- and 5-scroll attractors from a generalized Chua's circuit", *IEEE Trans. Circuits Syst. I,* vol. 47, pp. 425-429, 2000.

[66] M.E. Yalcin, J.A.K. Suykens, and J. Vandewalle, "On the realization of n-scroll attractors", in *Proc. IEEE Int. Symp. Circuits and Systems*, 1999, pp. 483-486.

[67] G. Zhong, K. F. Man, and G. Chen, "A systematic approach to generating n-scroll attractors", *Int. J. Bifurcat. Chaos;* vol. 12, pp. 2907-2915, 2002.

[68] S.M. Yu, S.S. Qiu, and Q.H. Lin, "New results of study on generating multiple-scroll chaotic attractors," *Science in China Series F;* vol. 46, pp. 104-115, 2003.

[69] K. S. Tang, G.Q. Zhong, G. Chen, and K.F. Man, "Generation of n-scroll attractors via sine function," *IEEE Trans. Circuits Syst. I;* vol. 48, pp. 1369-1372, 2001.

[70] S. Özoguz, A.S. Elwakil, and K. N. Salama, "n-Scroll chaos generator using nonlinear transconductor," *Electron. Lett.;* vol. 38, pp. 685-686, 2002.

[71] K.N. Salama, S. Özoguz, and A.S. Elwakil, "Generation of n-scroll chaos using nonlinear transconductors," in *Proc. IEEE Int. Symp. Circuits and Systems*, 2003, pp. 176-179.

[72] S. J. Linz, and J. C. Sprott, "Elementary chaotic flow," *Phys. Lett. A;* vol. 259, pp. 240-245, 1999.

[73] S.M. Yu, J. Lü, H. Leung, and G. Chen, "Design and circuit implementation of n-scroll chaotic attractor from a general Jerk system," *IEEE Trans. Circuits Syst. I;* vol. 52, pp. 1459-1476, 2005.

[74] S.M. Yu, J. Lü, H. Leung, and G. Chen, "N-scroll chaotic attractors from a general Jerk circuit", in *Proc. IEEE Int. Symp. Circuits and Systems*, 2005, pp. 1473-1476.

[75] M.E. Yalcin, J.A.K. Suykens, J. Vandewalle, and S. Özoguz, "Families of scroll grid attractors," *Int. J. Bifurcat. Chaos;* vol. 12, pp. 23-41, 2002.

[76] F. Han, J. Lü, X. Yu, G. Chen, and Y. Feng, "Generating multi-scroll chaotic attractors via a linear second-order hysteresis system," *Dyn. Contin. Discr. Impul. Syst. Series B*, vol. 12, pp. 95-110, 2005.

[77] J. Lü, F. Han, X. Yu and G. Chen, "Generating 3-D multi-scroll chaotic attractors: A hysteresis series switching method," *Automatica;* vol. 40, pp. 1677-1687, 2004.

[78] J. Lü, S.M. Yu, H. Leung and G. Chen, "Experimental verification for 3-D hysteresis multiscroll chaotic attractors," in *Proc. IEEE Int. Symp. Circuits and Systems*, 2005, pp. 3391-3394.

[79] J. Lü, G. Chen, X. Yu and H. Leung, "Design and analysis of multi-scroll chaotic attractors from saturated function series," *IEEE Trans. Circuits Syst. I*; vol. 51, pp. 2476-2490, 2004.

[80] J. Lü, T. Zhou, G. Chen and X. Yang, "Generating chaos with a switching piecewise-linear controller," *Chaos*, vol. 12, pp. 344-349, 2002.

[81] J. Lü, X. Yu and G. Chen, "Generating chaotic attractors with multiple merged basins of attraction: A switching piecewise-linear control approach," *IEEE Trans. Circuits Syst. I*; vol. 50, pp. 198-207, 2003.

[82] X. Yang and Q. Li, "Generate n-scroll attractors in linear system by scalar output feedback," *Chaos Solit. Fract.*, vol. 18, pp. 25-29, 2003.

[83] A. S. Elwakil, S. Özoguz and M.P. Kennedy, "Creation of a complex butterfly attractor using a novel Lorenz-type system," *IEEE Trans. Circuits Syst. I*, vol. 49, pp. 527-530, 2002.

[84] J. Lü, G. Chen and Y. Yu, "Asymptotic analysis of a modified Lorenz system," *Chin. Phys. Lett.*, vol. 19, pp. 1260-1263, 2002.

[85] R. Miranda and E. Stone, "The proto-Lorenz system," *Phys. Lett. A*, vol. 178, pp. 105-113, 1993.

[86] J. Lü, G. Chen and D.Z. Cheng, "A new chaotic system and beyond: The general Lorenz like system," *Int. J. Bifurcat. Chaos*, vol. 14, pp. 1507-1537, 2004.

[87] W.M. Ahmad, "Generation and control of multiscroll chaotic attractors in fractal order systems," *Chaos Solit. Fract.*, vol. 25, pp. 727–735, 2005.

[88] A.Y. Aguilar-Bustos, C. Cruz-Hernández, R.M. López-Gutiérrez, E. Tlelo-Cuautle, C. Posadas Castillo, "Hyperchaotic Encryption for Secure E-Mail Communication", in *Emergent Web Intelligence: Advanced Information Retrieval*. 1st ed. vol. XIX. R. Chbeir, Y. Badr, A. Abraham, A. Hassanien, (Eds.), London, UK: Springer, 2010, pp. 471-486.

[89] E.S.J. Martens, G.G.E. Gielen, *High-Level Modeling and Synthesis of Analog Integrated Systems*; Dordrecht, NL: Springer Science + Business Media; 2008.

[90] K. S. Kundert; *The Designer's Guide to Verilog-AMS*; Massachusetts, USA: Kluwer Academic Publishers; 2004.

[91] D. FitzPatrick, I. Miller; *Analog Behavioral Modeling with the Verilog-A Language*; Massachusetts, USA: Kluwer Academic Publishers Springer; 1997.

[92] R.A. Rutenbar, G. Gielen, J. Roychowdhury; "Hierarchical Modeling, Optimization, and Synthesis for System-Level Analog and RF Designs"; *Proceedings of the IEEE*, vol. 95, no.3, pp. 640-669, 2007.

[93] J. Roychowdhury, "Automated Macromodel generation for Electronic Systems", in *Proc. International Workshop on Behavioral Modeling and Simulation*, 2003, pp. 11-16.

[94] C. Sánchez-López, E. Tlelo-Cuautle, "Symbolic Behavioral Model Generation of Current-Mode Analog Circuits", in *Proc. IEEE Int. Symp. Circuits and Systems*, 2009, pp. 2761-2764.

[95] E. Tlelo-Cuautle, M.A. Duarte-Villaseñor, J.M. García-Ortega, C. Sánchez-López, "Designing SRCOs by combining SPICE and Verilog-A", *International Journal of Electronics*, vol. 94, no. 4, pp. 373-379, 2007.

[96] L.O. Chua, C.A. Desoer, E.S. Kuh, *Linear and Nonlinear Circuits*; New York, USA: McGraw-Hill; 1987.

[97] E. Yilmaz, G. Dundar, "Analog Layout Generator for CMOS Circuits", *IEEE Tran. Computer-Aided Design*, vol. 28, no. 1, pp. 32-45, 2009.

[98] M. Fakhfakh, E. Tlelo-Cuautle, F.V. Fernández, *Design of Analog Circuits through Symbolic Analysis*, Sharjah, U.A.E.:Bentham Sciences Publishers, 2010.

[99] T. Massier, H. Graeb, U. Schlichtmann, "The Sizing Rules Method for CMOS and Bipolar Analog Integrated Circuit Synthesis", *IEEE Tran. Computer-Aided Design*, vol. 27, no. 12, pp. 2209-2222, 2008.

[100] E. Martens, G. Gielen, "ANTIGONE: Top-down creation of analog-to-digital converter architectures", *Integration-the VLSI Journal*, vol. 42, no. 1, pp. 10-23, 2009.

[101] C. Sánchez-López, R. Trejo-Guerra, J. M. Muñoz-Pacheco, E. Tlelo-Cuautle, "N-scroll chaotic attractors from saturated functions employing CCII+s", *Nonlinear Dynamics*, 2010, in press, DOI 10.1007/s11071-009-9652-3.

[102] L.O. Chua, *Computer-Aided Analysis of Electronic Circuits: Algorithms and Computational Techniques*; New Jersey, USA: Prentice Hall; 1975.

[103] D. Leenaerts, W.M.G. van Bokhoven; *Piecewise Linear Modeling and Analysis*; Dordrecht, NL: Springer; 1998.

[104] W. K. Chen, J. Vandewalle and L. Vandenberghe L; *Piecewise-linear circuits and piecewise linear analysis: Circuits and Filters Handbook*; Florida, USA: CRC Press/IEEE Press; 1995.

[105] C. Cruz-Hernández, D. López, V. García, H. Serrano, R. Núñez, "Experimental realization of binary signals transmission using chaos", *J. Circuit. Syst. Comp.*, vol. 14, no. 3, pp. 453-468, 2005.

[106] http://www.analogdevices.org. [Accessed Nov. 18, 2009].

[107] http://focus.ti.com/docs/prod/folders/print/tl081.html. [Accessed Oct. 8, 2009].

[108] http://focus.ti.com/docs/prod/folders/print/tlc2262.html. [Accessed Oct. 23, 2009].

[109] M. G. R. Degrauwe, O. Nys, E. Dijkstra, J. Rijmenants, S. Bitz, B. L. A. G. Goffart, E. A. Vittoz, S. Cserveny, C. Meixenberger, G. van der Stappen, and H. J. Oguey, "IDAC: an interactive design tool for analog CMOS circuits," *IEEE J. Solid-State Circuits*, vol. 22, no. 6, pp. 1106-1116, 1987.

[110] R. Harjani, R. A. Rutenbar, and L. R. Carley, "OASYS: A Framework for Analog Circuit Synthesis," *IEEE Trans. Computer-Aided Design*, vol. 8, no. 12, pp. 1247-1265, 1989.

[111] F. El-Turky and E. E. Perry, "BLADES: an artificial intelligence approach to analog circuit design," *IEEE Trans. Computer-Aided Design*, vol. 8, no. 6, pp. 680-692, 1989.

[112] G. Jusuf, P. R. Gray, and A. L. Sangiovanni-Vincentelli, "CADICS-cyclic analog-to-digital converter synthesis," in *Proc. ACM/IEEE Int. Conf. on Computer-Aided Design*, 1990, pp. 286-289.

[113] G. Beenker, J. Conway, G. Schrooten, A. Slenter, "Analog CAD for consumer ICs," in *Analog circuit design,* J. Huijsing, R. van der Plassche and W. Sansen. Eds., Dordrecht, NL: Kluwer Academic Publishers, 1993, pp. 347- 367.

[114] J. D. Conway and G. G. Schrooten, "An automatic layout generator for analog circuits," in *Proc. European Design Automation Conf.*, 1992, pp. 513-519.

[115] B. R. Owen, R. Duncan, S. Jantzi, C. Ouslis, S. Rezania, and K. Martin, "BALLISTIC: an analog layout language," in *Proc. IEEE Custom Integrated Circuits Conf.*, 1995, pp. 41-44.

[116] E. Tlelo-Cuautle, I. Guerra-Gómez, C.A. Reyes-García, M.A. Duarte-Villaseñor, "Synthesis of Analog Circuits by Genetic Algorithms and their Optimization by Particle Swarm Optimization", in *Intelligent Systems for Automated Learning and Adaptation: Emerging Trends and Applications*, Raymond Chiong (Ed.), Pennsylvania, USA: IGI Global, 2010, pp. 173-192.

[117] F. Fernández, A. Rodríguez-Vázquez, J. L. Huertas, and G. Gielen; *Symbolic analysis techniques: applications to analog design automation.* New York, USA: IEEE Press, 1997.

[118] W. Nye, D. C. Riley, A. Sangiovanni-Vincentelli, and A. L. Tits, "DELIGHT.SPICE: an optimization-based system for the design of integrated circuits," *IEEE Trans. Computer-Aided Design*, vol. 7, no. 4, pp. 501-519, 1988.

[119] M. Krasnicki, R. Phelps, R. A. Rutenbar, and L. R. Carley, "MAELSTROM: efficient simulation-based synthesis for custom analog cells," in *Proc. ACM/IEEE Design Automation Conf.*, 1999, pp. 945-950.

[120] R. Phelps, M. J. Krasnicki, R. A. Rutenbar, L. R. Carley, and J. R. Hellums, "Anaconda: simulation-based synthesis of analog circuits via stochastic pattern search," *IEEE Trans. Computer- Aided Design*, vol. 19, no. 6, pp. 703-717, 2000.

[121] H. Y. Koh, C. H. Sequin, and P. R. Gray, "OPASYN: A compiler for CMOS operational amplifiers". *IEEE Tran. Computer-Aided Design*, vol. 9, no. 2, pp. 113-125, 1990.

[122] J. P. Harvey, M. I. Elmasry, and B. Leung, "STAIC: an interactive framework for synthesizing CMOS and BiCMOS analog circuits," *IEEE Trans. Computer-Aided Design*, vol. 11, no. 11, pp. 1402-1417, 1992.

[123] J. Rijmenants, J. B. Litsios, T. R. Schwarz, and M. G. R. Degrauwe, "ILAC: an automated layout tool for analog CMOS circuits," *IEEE J. Solid-State Circuits*, vol. 24, no. 2, pp. 417- 425, 1989.

[124] D. J. Garrod, R. A. Rutenbar, and L. R. Carley, "Automatic layout of custom analog cells in ANAGRAM," in *Proc. ACM/IEEE Int. Conf. on Computer-Aided Design*, 1998, pp. 544-547.

[125] E. Malavasi, U. Choudhury, and A. Sangiovanni-Vincentelli, "A routing methodology for analog integrated circuits," in *Proc. ACM/IEEE Int. Conf. on Computer-Aided Design*, 1990, pp. 202-205.

[126] E. Tlelo-Cuautle, J.M. Muñoz-Pacheco, "Numerical simulation of Chua´s circuit oriented to circuit synthesis", *Int. J. Nonlinear Sci.*, vol. 8, no. 2, pp. 249-256, 2007.

[127] E. Tlelo-Cuautle, J.M. Muñoz-Pacheco. "Automatic Simulation of 1D and 2D Chaotic Oscillators", *Journal of Physics: Conference Series*, vol. 96 012059, pp.1-10, 2008.

[128] J.M. Muñoz-Pacheco, E. Tlelo-Cuautle, "Synthesis of n-Scrolls Attractors using Saturated Functions from High-Level Simulation". *Journal of Physics: Conference Series*, vol. 96 012050, pp. 1-8, 2008.

[129] E. Tlelo-Cuautle, J.M. Muñoz-Pacheco, "Simulation of Chua's circuit by automatic control of step-size", *Applied Mathematics and Computation*, vol. 190, issue 2, pp. 1526-1533, 2007.

[130] E. Tlelo-Cuautle, J.M. Muñoz-Pacheco, J. Martínez-Carballido, "Frequency-scaling simulation of Chua's circuit by automatic determination and control of step-size", *Applied Mathematics and Computation*, vol. 194, issue 2, pp. 486-491, 2007.

[131] K. Ramasubramanian, M.S. Sriram, "A comparative study of computation of Lyapunov spectra with different algorithms", *Physica D*, vol. 139, pp. 72-86, 2000.

[132] J. Lu, G. Yang, H. Oh, A.C.J. Luo, "Chaos Computing Lyapunov exponents of continuous dynamical systems: method of Lyapunov vectors", *Chaos Solit. Fract.*, vol. 23, pp. 1879-1892, 2005.

[133] L. Dieci, "Jacobian Free Computation of Lyapunov Exponents", *Journal of Dynamics and Differential Equations*, vol. 14-3, pp. 697-717, 2002.

[134] J.M. Muñoz-Pacheco, E. Tlelo-Cuatle, "Automatic Synthesis of 2D-n-scrolls Chaotic Systems by Behavioral Modeling", *Journal of Applied Research and Technology*. vol. 7, no. 1, pp. 5-14, 2009.

[135] J.M. Muñoz-Pacheco, E. Tlelo-Cuautle, V.H. Carbajal-Gómez, "A CAD-Tool for the Design of n-Scrolls Chaotic Systems from Behavioral Modeling", in *Proc. Second International Workshop on Nonlinear Dynamics and Synchronization*, 2009, pp.198-202.

[136] J.M. Muñoz-Pacheco, "Synthesis of Chaotic Oscillators by Applying Behavioral Modeling", Ph.D. thesis, National Institute for Astrophysics, Optics and Electronics (INAOE), Puebla, Mexico, August 14, 2009.

INDEX

www.ingramcontent.com/pod-product-compliance
Lightning Source LLC
Chambersburg PA
CBHW041722210326
41598CB00007B/743